Wireless Networks

Series Editor

Xuemin Sherman Shen, University of Waterloo, Waterloo, ON, Canada

The purpose of Springer's Wireless Networks book series is to establish the state of the art and set the course for future research and development in wireless communication networks. The scope of this series includes not only all aspects of wireless networks (including cellular networks, WiFi, sensor networks, and vehicular networks), but related areas such as cloud computing and big data. The series serves as a central source of references for wireless networks research and development. It aims to publish thorough and cohesive overviews on specific topics in wireless networks, as well as works that are larger in scope than survey articles and that contain more detailed background information. The series also provides coverage of advanced and timely topics worthy of monographs, contributed volumes, textbooks and handbooks.

** Indexing: Wireless Networks is indexed in EBSCO databases and DPLB **

More information about this series at http://www.springer.com/series/14180

Jie Gao • Mushu Li • Weihua Zhuang

Connectivity and Edge Computing in IoT: Customized Designs and AI-based Solutions

 Springer

Jie Gao
Department of Electrical and Computer
Engineering
Marquette University
Milwaukee, WI, USA

Mushu Li
Department of Electrical and Computer
Engineering
University of Waterloo
Waterloo, ON, Canada

Weihua Zhuang
Department of Electrical and Computer
Engineering
University of Waterloo
Waterloo, ON, Canada

ISSN 2366-1186   ISSN 2366-1445   (electronic)
Wireless Networks
ISBN 978-3-030-88745-2   ISBN 978-3-030-88743-8   (eBook)
https://doi.org/10.1007/978-3-030-88743-8

This Springer imprint is published by the registered company Springer Nature Switzerland AG
The registered company address is: Gewerbestrasse 11, 6330 Cham, Switzerland

Preface

The Internet of Things (IoT) is revolutionizing the world and impacting the daily lives of billions of people. Supporting use cases for households, manufacturers, transportation, agriculture, healthcare, and much more, IoT carries many potentials and expectations for prospering human society. Technologically, we are at an early stage of IoT development, aiming at connecting tens of billions of devices to make homes, communities, factories, farms, and everywhere else smart and automated. Tremendous efforts are necessary to advance IoT research and development.

Two cornerstones of IoT are data collection/exchange and data analysis. The former demands connectivity solutions, while the latter requires computing solutions. Due to the broad scope of IoT and the drastically different characteristics and requirements of IoT use cases, no "one-size-fits-all" design can meet the expectations of all use cases. Therefore, customizing connectivity or computing solutions for specific use cases is challenging yet essential. There are many system features and performance measures to consider in the customization, such as connection link density, resource overhead, transmission and computation delay, service reliability, energy efficiency, and device mobility, and making proper trade-offs among them is critical.

Accounting for all performance metrics and making optimal trade-offs can yield high complexity. Correspondingly, artificial intelligence (AI) solutions, such as neural networks and reinforcement learning, can become useful. Powered by AI methods, connectivity or computing solutions can learn from experience to handle the complexity, assuming that sufficient data are available for training. Specifically, AI can play various roles in IoT, including data traffic load prediction, access control, and computation task scheduling, to name a few.

In this book, we focus on connectivity and edge computing in IoT and present our designs for four representative IoT use cases, i.e., smart factory, rural IoT, Internet of vehicles, and mobile virtual reality. We thoroughly review the existing research in this field, including many works published in recent years. Then, through innovative designs, we demonstrate the necessity and potential of customizing solutions based on the use cases. In addition, we exploit AI methods to empower our solutions. The four research works included in this book serve a collective objective:

enabling on-demand data collection and/or analysis for IoT use cases, especially in resource-limited IoT systems. We hope that this book will inspire further research on connectivity and edge computing in the field of IoT.

Milwaukee, WI, USA   Jie Gao

Waterloo, ON, Canada   Mushu Li

Waterloo, ON, Canada   Weihua Zhuang

July 2021

Acknowledgements

The authors would like to thank Professor Xuemin (Sherman) Shen at the University of Waterloo, the Editor of the *Wireless Networks* series, for his support in publishing this book and his comments and suggestions on its technical content.

We would also like to thank Professor Lian Zhao at the Ryerson University for her helpful discussions on the research presented in Chaps. 3 and 4.

In addition, we thank Dr. Xu Li, Professor Nan Cheng, and Conghao Zhou, who participated in the research discussed in Chaps. 2, 4, and 5, respectively.

We appreciate valuable discussions with the members of the Broadband Communications Research (BBCR) Lab at the University of Waterloo.

Special thanks to Susan Lagerstrom-Fife, Senior Editor at Springer New York, and Shina Harshavardhan, Project Coordinator for Springer Nature for their help during the preparation of this monograph.

Contents

Acronyms

3GPP	Third generation partnership project
5G	Fifth generation
5G NR	5G new radio
AC	Actor-critic
AD	Access delay
AD-F	Access delay counted in frames
ADMM	Alternating direction method of multipliers
AI	Artificial intelligence
AP	Access point
BFoV	Base field-of-view
BS	Base station
CNN	Convolutional neural network
CSMA	Carrier-sense multiple access
DCF	Distributed coordination function
DDPG	Deep deterministic policy gradient
DNN	Deep neural network
DQN	Deep Q network
DRL	Deep reinforcement learning
EDT	Early data transmission
EFoV	Extended field-of-view
eMBB	Enhanced mobile broadband
ET	Enhanced tile
FoV	Field-of-view
HMD	Head-mounted device
HP	High priority
IIoT	Industrial Internet of Things
IoT	Internet of Things
IoV	Internet of Vehicles
IT	Information technology
LoRa	Long range
LP	Low priority

LPWA	Low-power wide-area
LSTM	Long short-term memory
LTE	Long-term evolution
LTE-M	Long-term evolution for machine-type communications
M2M	Machine-to-machine
MAC	Medium access control
MDP	Markov decision process
mMTC	Massive machine-type communications
mmWave	Millimeter-wave
MsCS	Mini-slot based carrier sensing
MSE	Mean squared error
MTC	Machine-type communication
NB-IoT	Narrowband IoT
NOMA	Non-orthogonal multiple access
QoE	Quality of experience
QoS	Quality of service
RACH	Random access channel
RAW	Restricted access window
RMAB	Restless multi-armed bandit
RP	Regular priority
RSU	Roadside unit
SCA	Successive convex approximation
SOC	Second order cone
SMsA	Superimposed mini-slot assignment
SyncCS	Synchronization carrier sensing
TDMA	Time-division multiple access
TPSA	Task partition and scheduling algorithm
TTI	Transmission time interval
UAV	Unmanned aerial vehicle
URLLC	Ultra-reliable low-latency communications
V2I	Vehicle-to-infrastructure
V2X	Vehicle-to-everything
VR	Virtual reality
VS	Video segment
WI	Whittle index
WLAN	Wireless local area network

Chapter 1
Introduction

In this chapter, we first provide an overview of the Internet of Things from the perspectives of connected devices, use cases, deployment efforts, and technical advancement. Then, connectivity and edge computing in IoT are introduced, respectively, focusing on the requirements, available options, and challenges. The role of artificial intelligence in IoT and challenges in developing AI-based solutions are also discussed. Last, we present the scope and organization of this book.

1.1 The Era of Internet of Things

We are entering the era of the Internet of Things (IoT). Targeting to connect billions of devices, such as wearables, appliances, and industrial actuators, and a variety of systems, such as sensor networks, transportation management centers, and power grids, IoT has become a major driver worldwide for innovations in both business and technology development. The global IoT market size in 2020 is estimated to be approximately 309 billions in USD, and the forecast for 2021 and 2028 is 831 billions and 1855 billions, respectively, with an annual growth rate of 25.4% between 2021 and 2028 [1]. Meanwhile, the number of networked devices is expected to increase from around 20 billions in 2020 to almost 30 billions in 2023, with almost 15 billion machine-to-machine (M2M) connections in 2023 [2]. Moreover, it is predicted that platforms connecting devices, cloud servers, and application providers will harvest comparable revenue from emerging IoT use cases and from traditional information technology (IT) use cases by 2023 [3].

IoT is a broad concept that covers a wide range of use cases. In manufacturing industries, IoT solutions can improve asset management, optimize supply chains, and enable factory automation [4]. In agriculture, IoT platforms can facilitate plant status monitoring, and pest and disease control [5]. In urban management, IoT techniques can enable smart cities by integrating smart street lighting, intelligent traffic control, fire and pollution detection, etc., to promote safe, comfortable, and energy-efficient living conditions [6]. In healthcare, IoT applications can support remote in-home health monitoring for proactive and preventive diagnosis interventions [7]. In the airline business, IoT platforms can reduce fuel costs and service disruption and thereby improve customer experience [8]. Other promising IoT applications include crude oil production, wildfire detection, search and rescue, smart campus, augmented shopping, and so on [9–13].

Many countries and regions have started IoT programs or pilot projects. For example, the IoT European Large-Scale Pilots Programme has been promoting partnerships across Europe since 2016 and conducting various IoT projects with a total budget of €100 millions, including ACTIVAGE (for elderly smart living), AUTOPILOT (for automated driving), IoF2020 (for the Internet of food and farm) [14]. In the United States, New York City published its IoT strategy in March 2021 with an objective to create an IoT ecosystem for consumer, industry, and government use cases [15], while other cities, such as Las Vegas, are on course to become smart cities [16]. In China, the number of licensed IoT connections has reached 600 millions by 2018, and a major focus of future IoT development is intelligent manufacturing [17]. In addition to the above programs or pilot projects, many industry leaders have invested in and developed IoT platforms, examples of which include Amazon Web Services, Microsoft Azure IoT Platform, IBM Watson IoT Platform, and Siemens MindSphere [18].

Besides various investment from governments and industries, technology advancement in device hardware, software, communications, cloud/edge computing, artificial intelligence (AI), etc., have been propelling the development and deployment of IoT. Improvement in hardware enables the production of IoT devices with smaller sizes and lower costs [19]. Improvements in software allow IoT devices and platforms to become more secure, reliable, and energy-efficient [20, 21]. Advancement in communication technologies enables a massive scale of connections required for realizing IoT as well as new communication paradigms, such as M2M communications [22, 23]. Modern cloud and edge computing technologies provide versatile paradigms of data processing for IoT applications, allowing on-demand computing service provisioning through task offloading [24]. Lastly, advances in AI techniques render intelligent and automated connectivity and computing solutions in IoT [25].

This book focuses on the connectivity and computing aspects of IoT, with a particular focus on use case-specific designs and AI-based solutions. The rest of this chapter will discuss the basics of connectivity, edge computing, and the role of AI in IoT.

1.2 Connectivity in IoT

Connectivity is the foundation of IoT as it enables data collection from or exchange among networked IoT devices. Different network topology, connectivity requirements, and connectivity options may apply in IoT, depending on the application.

Regarding network operation, IoT applications can be implemented in a distributed, a decentralized, or a centralized manner. Examples of distributed IoT applications include distributed sensing and communication in autonomous driving [26] and plant monitoring for predictive maintenance in manufacturing [27], which require a low response time and need to process collected data locally. Examples of decentralized IoT applications include localization [28] and edge computing [29], which leverage infrastructure and resources on network edge to serve end users without relying on cloud servers. Additionally, many IoT applications adopt a client-server mode and exploit centralized cloud computing platforms, such as the enterprise platforms mentioned in Sect. 1.1. Examples of such applications are metropolitan-area intelligent transportation system planning [30] and large-scale supervisory control and data acquisition [31], which rely on extensive computing and storage resources provided by data centers.

Regarding connectivity requirements, IoT applications may require high connection density, low communication delay, long communication range, high transmission rate, or combinations of those. In a smart city scenario, 30,000 connections per square kilometer (km^2) may be needed just for connecting household water, electricity, and gas meters, which send messages with intervals between 30 min and 24 h [32]. Such connections are delay-tolerant and usually have a short range, e.g., 15 meters (m). In a factory automation setting, process state monitoring may involve 10,000 devices per km^2 [33]. Such connections span factory plants with a typical size around 300 m × 300 m × 50 m, and the delay tolerance is on the level of 50 milliseconds (ms). In internet of vehicles (IoV), a vehicle may need to simultaneously communicate with hundreds of other vehicles [34]. The connections for such communications can be transient, and the delay tolerance can be very strict, e.g., 10 ms for road safety applications.

Regarding connectivity options, various wireless communication standards and techniques are available for IoT. The three use cases of the fifth generation (5G) cellular networks, i.e., enhanced mobile broadband (eMBB), ultra-reliable low-latency communications (URLLC), and massive machine-type communications (mMTC), aim at providing support for various IoT applications [35]. Meanwhile, 802.11ax, or Wi-Fi 6, has enhanced support for IoT and is suitable for smart home applications [36]. In addition, a few low-power wide-area (LPWA) technologies and standards, such as Long-Term Evolution for Machine-Type Communications (LTE-M), Long Range (LoRa), Narrowband IoT (NB-IoT), support cost-effective long-range communications and are suitable for applications such as smart logistics and environment or wild-life monitoring [37]. In the future, IoT devices may also be connected via satellites.

Given the varieties of IoT applications and their connectivity requirements, finding optimal connectivity solutions is challenging, and such challenge is aggrandized when considering network heterogeneity, device mobility, network resource limitations, cost-effectiveness, and scalability. As a result, despite various potential options as mentioned above, customized designs are necessary for providing the best support to specific applications due to their unique characteristics and requirements. In Chap. 2, we will customize a connectivity solution for industrial IoT and demonstrate the potential of such customized designs for connecting IoT devices. In Chaps. 3 and 4, we will present connectivity solutions related to computing task offloading and result delivery in edge computing.

1.3 Edge Computing in IoT

Most IoT applications require not only data collection or exchange but also data analysis. As a result, they demand a computing paradigm and related resources. The data processing may happen on end user devices (such as sensors or vehicles), edge facilities (such as local network controllers), or cloud computing servers (such as Amazon Elastic Compute Cloud).

On-device processing is feasible for devices such as smartphones and vehicles, which have the hardware, software, and other resources for on-board computing [38]. Meanwhile, a significant portion of IoT devices, such as sensors and parking meters, are low-cost devices with limited processing power, storage, or battery [39]. With no or minimum on-device processing capability, such devices may resort to cloud computing and leverage resources in a cloud for data processing [40]. The cloud computing paradigm enables a variety of IoT applications and is especially suitable for applications running in a client-server mode. However, cloud computing requires devices to upload the data for processing to a cloud server, which can cause excessive traffic loads for the IoT networks when a massive number of devices rely on cloud computing. In addition, the round-trip communication, i.e., data uploading and computing result delivery, can cause a large delay that is unacceptable for applications such as autonomous driving and industrial robot arm control [41]. To reduce network traffic load and delay, edge computing has emerged as a solution, in which computing resources are deployed outside of the cloud and close to end users on network edge [42]. Such a computing paradigm is known as mobile edge computing or multi-access edge computing (MEC).

With the advent of edge computing, applications that require low-delay computing can leverage computing servers on the network edge [43]. This creates new opportunities for both IoT service providers and network operators. In smart healthcare, data collected by smartphones or wearable devices can be processed at an edge server for health monitoring applications such as gait analysis and fall risk assessment [44]. In smart cities, videos captured by cameras can be processed at edge servers for surveillance and event recognition [45, 46]. In autonomous driving, vehicles can upload data collected by cameras, radars, and other sensors to edge servers and enhance road safety via data analysis such as object recognition and tracking. In addition, many applications in various domains that leverage edge computing are emerging [47].

On the other hand, edge computing renders IoT networks more complex. New challenges arise, which often involve the synergy of computing and connectivity. For example, edge computing servers can be deployed at the access points (APs) of femtocells (e.g., home networks), small cells, and macrocells, and each deployment option has its own pros and cons [48]. In addition, the joint scheduling of transmission and computing tasks becomes critical for supporting applications with stringent delay requirements [49]. In highly dynamic networks such as vehicular networks, computing service migration or collaborative computing can be necessary for handling device mobility [50]. In Chaps. 3–5, we present edge computing solutions in representative IoT scenarios, such as IoV, and discuss various issues related to edge computing, such as task scheduling, content caching, collaborative computing, and computing result delivery.

1.4 AI in IoT

The world has witnessed a rapid advancement of AI in the past decade, with many successful real-world applications, especially in the field of natural language processing and computer vision [51]. Such success inspires the investigation on potential applications of AI in IoT, and many ideas have emerged for various use cases, such as mining, healthcare, and transportation [52, 53].

Incorporating of AI in IoT is natural. First, involving a massive number of devices, diverse applications, and spatiotemporally-variant service demands, IoT networks are complex and dynamic. AI potentially offers a viable alternative approach to managing IoT networks with the desired scalability and adaptability, while satisfying diverse and often stringent application requirements. Second, the effectiveness of AI relies on abundant data, e.g., for training neural networks, while a massive number of IoT devices can generate or provide a massive amount of data to fuel AI. Last, AI methods are suitable for data analysis in many IoT applications, such as health monitoring and fault pattern identification in smart grids [54].

AI can play a multifarious role in IoT, in terms of both the connectivity and the edge computing. Specifically, AI can be used for network traffic load prediction to facilitate IoT network planning [55]. AI can also be adopted in medium access control (MAC) to enhance IoT network throughput or fairness [56]. In addition, AI can be applied to handle computing task scheduling [57], offloading [58], and migration [59] for effective edge computing with minimum computing delay, balanced computing load distribution, or adaptivity to network dynamics.

Despite a tremendous potential of AI in empowering various IoT applications, many challenges exist in AI-based solutions for IoT. Specifically, choosing appropriate AI methods for considered IoT applications, while taking practicality into account, is essential yet challenging. Moreover, AI functionality deployment, communication overhead, data processing delay, and scalability of AI-based solutions, among other possible issues, all need to be accounted for. In Chaps. 2, 4, and 5, we develop AI-based connectivity and edge computing solutions in representative IoT scenarios, including learning assisted scheduling, collaborative computing, and content distribution. These solutions demonstrate the potentials and advantages of incorporating AI into various IoT applications.

1.5   Scope and Organization of This Book

In this book, we focus on the connectivity and edge computing aspects of IoT. We develop customized designs and AI-based solutions for connectivity and/or edge computing in representative IoT use cases, including smart factory, rural IoT, IoV, and mobile virtual reality (VR).

In Chap. 2, we investigate MAC for an industrial IoT network. Considering a local area network with high device density, short packets, and stringent delay and reliability requirements, we tailor a MAC protocol for smart factory applications and design a neural network to assist the scheduling of transmission opportunities for industrial IoT devices.

In Chap. 3, we investigate unmanned aerial vehicle (UAV) assisted edge computing for rural IoT applications such as in smart agriculture or forest monitoring. Using a UAV to provide connectivity and computing service to IoT devices, we develop a solution to jointly optimize the connectivity, through determining the UAV trajectory and device transmit power, and the edge computing, through properly allocating computing load between the UAV and the devices.

In Chap. 4, we investigate edge computing for delay-sensitive applications in IoV to improve the safety or driving experience of drivers. To address the challenge of high vehicle mobility, we adopt collaborative edge computing to reduce computing delay and improve computing service reliability for vehicles and develop a deep reinforcement learning assisted approach to find the optimal computing task offloading and computing result delivery policy.

In Chap. 5, we investigate edge-assisted content caching and distribution for mobile VR video streaming, which requires edge computing to render some VR videos. To improve the viewer's quality of experience (QoE), we design a scheme to cache video content and reduce frame missing in VR video streaming, and develop a deep reinforcement learning based scheme for scheduling VR content delivery to viewers.

In Chap. 6, we conclude this book and briefly discuss further research directions in connectivity and edge computing in IoT.

References

1. Internet of Things (IoT) market size, share & covid-19 impact analysis, by component (platform, solution & services), by end use industry (BFSI, retail, government, healthcare, manufacturing, agriculture, sustainable energy, transportation, IT & telecom, others), and regional forecast, 2021-2028. Tech. Rep. Report ID: FBI100307, Fortune Business Insights
2. Cisco annual Internet report (2018–2023), white paper. Tech. rep., Cisco
3. Dahlqvist, F., Patel, M., Rajko, A., Shulman, J.: Growing opportunities in the Internet of Things. Tech. rep., McKinsey (2019)
4. Yang, C., Shen, W., Wang, X.: The Internet of Things in manufacturing: key issues and potential applications. IEEE Syst. Man Cybern. Mag. 4(1), 6–15 (2018)
5. Elijah, O., Rahman, T.A., Orikumhi, I., Leow, C.Y., Hindia, M.N.: An overview of Internet of Things (IoT) and data analytics in agriculture: benefits and challenges. IEEE Internet Things J. 5(5), 3758–3773 (2018)
6. Du, R., Santi, P., Xiao, M., Vasilakos, A.V., Fischione, C.: The sensable city: a survey on the deployment and management for smart city monitoring. IEEE Commun. Surv. Tut. 21(2), 1533–1560 (2019)
7. Philip, N.Y., Rodrigues, J.J.P.C., Wang, H., Fong, S.J., Chen, J.: Internet of Things for in-home health monitoring systems: current advances, challenges and future directions. IEEE J. Sel. Areas Commun. 39(2), 300–310 (2021)
8. Lavoie-Tremblay, K., Gautam, S., Levine, G.: Connecting the dots on IoT for the industrial world. IEEE Internet Things Mag. 1(1), 24–26 (2018)
9. Duan, Q., Sun, D., Li, G., Yang, G., Yan, W.W.: IoT-enabled service for crude-oil production systems against unpredictable disturbance. IEEE Trans. Services Comput. 13(4), 759–768 (2020)
10. Bushnaq, O.M., Chaaban, A., Al-Naffouri, T.Y.: The role of UAV-IoT networks in future wildfire detection. IEEE Internet Things J. (2021). https://doi.org/10.1109/JIOT.2021.3077593
11. Bianco, G.M., Giuliano, R., Marrocco, G., Mazzenga, F., Mejia-Aguilar, A.: LoRa system for search and rescue: path-loss models and procedures in mountain scenarios. IEEE Internet Things J. 8(3), 1985–1999 (2021)
12. Sutjarittham, T., Habibi Gharakheili, H., Kanhere, S.S., Sivaraman, V.: Experiences with IoT and AI in a smart campus for optimizing classroom usage. IEEE Internet Things J. 6(5), 7595–7607 (2019)
13. Hormann, L.B., Putz, V., Rudic, B., Kastl, C., Klinglmayr, J., Pournaras, E.: Augmented shopping experience for sustainable consumption using the Internet of Thing. IEEE Internet Things Mag. 2(3), 46–51 (2019)
14. IoT European large-scale pilots programme - large scale pilots projects. IoT European Large-Scale Pilots Programme (2018). https://european-iot-pilots.eu/wp-content/uploads/2018/03/220315_SD_IoT_Brochure_A4_LowRes_final-1.pdf
15. The New York City Internet of Things strategy. version 1.26.0402 (2021). https://www1.nyc.gov/assets/cto/downloads/iot-strategy/nyc_iot_strategy.pdf
16. Dickens, C., Boynton, P., Rhee, S.: Principles for designed-in security and privacy for smart cities. In: Proceedings of the 4th Workshop Int. Sci. Smart City Operations Platforms Eng. (SCOPE), pp. 25–29 (2019)
17. Giaffreda, R.: IoT in China: What does the future hold? IEEE Internet Things Mag. 2(3), 52–53 (2019)
18. Hoffmann, J.B., Heimes, P., Senel, S.: IoT platforms for the Internet of Production. IEEE Internet Things J. 6(3), 4098–4105 (2019)
19. Folea, S.C., Mois, G.D.: Lessons learned from the development of wireless environmental sensors. IEEE Trans. Instrum. Meas. 69(6), 3470–3480 (2020)
20. Atlam, H.F., Wills, G.B.: IoT security, privacy, safety and ethics. In: Digital Twin Technologies and Smart Cities, pp. 123–149. Springer, Cham (2020)

21. Georgiou, K., Xavier-de Souza, S., Eder, K.: The IoT energy challenge: a software perspective. IEEE Embedded Syst. Lett. **10**(3), 53–56 (2018)

22. Gazis, V.: A survey of standards for machine-to-machine and the Internet of Things. IEEE Commun. Surv. Tut. **19**(1), 482–511 (2017)

23. Gao, J., Li, M., Zhao, L., Shen, X.: Contention intensity based distributed coordination for V2V safety message broadcast. IEEE Trans. Veh. Technol. **67**(12), 12,288–12,301 (2018)

24. Porambage, P., Okwuibe, J., Liyanage, M., Ylianttila, M., Taleb, T.: Survey on multi-access edge computing for Internet of Things realization. IEEE Commun. Surv. Tut. **20**(4), 2961–2991 (2018)

25. Javaid, N., Sher, A., Nasir, H., Guizani, N.: Intelligence in IoT-based 5G networks: opportunities and challenges. IEEE Commun. Mag. **56**(10), 94–100 (2018)

26. Philip, B.V., Alpcan, T., Jin, J., Palaniswami, M.: Distributed real-time IoT for autonomous vehicles. IEEE Trans. Ind. Inf. **15**(2), 1131–1140 (2019)

27. Liu, Y., Yu, W., Dillon, T.S., Rahayu, W., Li, M.: Empowering IoT predictive maintenance solutions with AI: a distributed system for manufacturing plant-wide monitoring. In: IEEE Transactions on Industrial Informatics (2021)

28. Kasmi, Z., Guerchali, N., Norrdine, A., Schiller, J.H.: Algorithms and position optimization for a decentralized localization platform based on resource-constrained devices. IEEE Trans. Mobile Comput. **18**(8), 1731–1744 (2019)

29. Cicconetti, C., Conti, M., Passarella, A.: A decentralized framework for serverless edge computing in the Internet of Things. IEEE Trans. Netw. Service Manag. **18**(2), 2166–2180 (2021)

30. Lidkea, V.M., Muresan, R., Al-Dweik, A.: Convolutional neural network framework for encrypted image classification in cloud-based ITS. IEEE Open J. Intell. Transp. Syst. **1**, 35–50 (2020)

31. Sajid, A., Abbas, H., Saleem, K.: Cloud-assisted IoT-based scada systems security: a review of the state of the art and future challenges. IEEE Access **4**, 1375–1384 (2016)

32. Ericsson mobility report - on the pulse of the networked society. Tech. rep., Ericsson (2016)

33. Technical specification group services and system aspects; service requirements for the 5G system; stage 1 (Release 16). Tech. Rep. TS 22.261, Version 16.14.0, 3GPP (2021)

34. Bi, Y., Zhou, H., Zhuang, W., Zhao, H.: Safety Message Broadcast in Vehicular Networks. Springer, Cham (2017)

35. Li, J., Shi, W., Yang, P., Ye, Q., Shen, X.S., Li, X., Rao, J.: A hierarchical soft RAN slicing framework for differentiated service provisioning. IEEE Wireless Commun. **27**(6), 90–97 (2020)

36. Wi-Fi 6 industry impact report - transition or transformation? Tech. rep., FeibusTech

37. Raza, U., Kulkarni, P., Sooriyabandara, M.: Low power wide area networks: an overview. IEEE Commun. Surv. Tut. **19**(2), 855–873 (2017)

38. Rawassizadeh, R., Pierson, T.J., Peterson, R., Kotz, D.: NoCloud: exploring network disconnection through on-device data analysis. IEEE Pervasive Comput. **17**(1), 64–74 (2018)

39. Ciuffoletti, A.: Low-cost IoT: a holistic approach. J. Sens. Actuator Netw. **7**(2), 19 (2018)

40. Barcelo, M., Correa, A., Llorca, J., Tulino, A.M., Vicario, J.L., Morell, A.: IoT-cloud service optimization in next generation smart environments. IEEE J. Sel. Areas Commun. **34**(12), 4077–4090 (2016)

41. Zhang, B., Mor, N., Kolb, J., Chan, D.S., Lutz, K., Allman, E., Wawrzynek, J., Lee, E., Kubiatowicz, J.: The cloud is not enough: Saving IoT from the cloud. In: Proceedings of the 7th USENIX Workshop Hot Topics Cloud Comput. (HotCloud 15). USENIX Association, Santa Clara, CA (2015)

42. Shi, W., Cao, J., Zhang, Q., Li, Y., Xu, L.: Edge computing: vision and challenges. IEEE Internet Things J. **3**(5), 637–646 (2016)

43. Ashouri, M., Davidsson, P., Spalazzese, R.: Cloud, edge, or both? Towards decision support for designing IoT applications. In: Proc. Fifth Int. Conf. Internet Things: Syst., Manage. Secur., pp. 155–162 (2018)

44. Baktir, A.C., Tunca, C., Ozgovde, A., Salur, G., Ersoy, C.: SDN-based multi-tier computing and communication architecture for pervasive healthcare. IEEE Access **6**, 56765–56781 (2018)
45. Long, C., Cao, Y., Jiang, T., Zhang, Q.: Edge computing framework for cooperative video processing in multimedia IoT systems. IEEE Trans. Multimedia **20**(5), 1126–1139 (2018)
46. Liu, S., Liu, L., Tang, J., Yu, B., Wang, Y., Shi, W.: Edge computing for autonomous driving: opportunities and challenges. Proc. IEEE **107**(8), 1697–1716 (2019)
47. Porambage, P., Okwuibe, J., Liyanage, M., Ylianttila, M., Taleb, T.: Survey on multi-access edge computing for Internet of Things realization. IEEE Commun. Surv. Tut. **20**(4), 2961–2991 (2018)
48. Vallati, C., Virdis, A., Mingozzi, E., Stea, G.: Mobile-edge computing come home connecting things in future smart homes using LTE device-to-device communications. IEEE Consum. Electron. Mag. **5**(4), 77–83 (2016)
49. He, H., Shan, H., Huang, A., Ye, Q., Zhuang, W.: Edge-aided computing and transmission scheduling for LTE-U-enabled IoT. IEEE Trans. Wireless Commun. **19**(12), 7881–7896 (2020)
50. Wang, S., Guo, Y., Zhang, N., Yang, P., Zhou, A., Shen, X.: Delay-aware microservice coordination in mobile edge computing: a reinforcement learning approach. IEEE Trans. Mobile Comput. **20**(3), 939–951 (2021)
51. Wiriyathammabhum, P., Summers-Stay, D., Fermüller, C., Aloimonos, Y.: Computer vision and natural language processing: recent approaches in multimedia and robotics. ACM Comput. Surv. **49**(4), 1–44 (2016)
52. Al-Turjman, F.: Artificial Intelligence in IoT. Springer, Cham (2019)
53. Manoharan, K.G., Nehru, J.A., Balasubramanian, S.: Artificial Intelligence and IoT: Smart Convergence for Eco-friendly Topography, vol. 85. Springer Nature, Singapore (2021)
54. Bose, B.K.: Artificial intelligence techniques in smart grid and renewable energy systems–some example applications. Proc. IEEE **105**(11), 2262–2273 (2017)
55. Wang, W., Zhou, C., He, H., Wu, W., Zhuang, W., Shen, X.: Cellular traffic load prediction with LSTM and Gaussian process regression. In: Proceedings of the 2020 IEEE International Conference Communication (ICC), pp. 1–6 (2020)
56. Yu, Y., Wang, T., Liew, S.C.: Deep-reinforcement learning multiple access for heterogeneous wireless networks. IEEE J. Sel. Areas Commun. **37**(6), 1277–1290 (2019)
57. Zhou, C., Wu, W., He, H., Yang, P., Lyu, F., Cheng, N., Shen, X.: Deep reinforcement learning for delay-oriented IoT task scheduling in SAGIN. IEEE Trans. Wireless Commun. **20**(2), 911–925 (2021)
58. Ye, Q., Shi, W., Qu, K., He, H., Zhuang, W., Shen, X.: Joint RAN slicing and computation offloading for autonomous vehicular networks: a learning-assisted hierarchical approach. IEEE Open J. Veh. Technol. **2**, 272–288 (2021)
59. Yuan, Q., Li, J., Zhou, H., Lin, T., Luo, G., Shen, X.: A joint service migration and mobility optimization approach for vehicular edge computing. IEEE Trans. Veh. Technol. **69**(8), 9041–9052 (2020)

Chapter 2
Industrial Internet of Things: Smart Factory

In this chapter, we investigate the smart factory use case in the scenario of industrial IoT, focusing on the connectivity aspect. First, through reviewing the connectivity requirements and related standards, we illustrate the insufficiency of existing techniques in meeting the expectations of smart factories and the necessity of tailoring connectivity solutions. Then, we design a novel medium access control protocol, which features grant-free distributed channel access, to support high device density and low communication latency with low communication overhead. We further propose a deep neural network-assisted centralized approach to configure the protocol parameters and schedule transmission opportunities for all devices. Combining the customized protocol and the AI-assisted scheduling, our design demonstrates promising potentials for smart factories by simultaneously enabling massive connections and millisecond-level delay for high priority devices.

2.1 Industrial IoT Networks

Industrial Internet of Things (IIoT) utilizes connected devices, including sensors, actuators, and controllers to automate data collection and analysis for increasing productivity, reducing energy consumption, and improving safety and reliability in various industries, such as manufacturing, construction, warehouses, and oil rigs and refineries [1, 2]. As one of the most promising technology domains, IIoT is envisioned to reshape industries around the world, e.g., creating "factories of the future". As a result, IIoT related research and development are attracting widespread attention, and the global IIoT market is expected to reach 263.4 billions in US dollars by 2027 [3].

© The Author(s), under exclusive license to Springer Nature Switzerland AG 2021
J. Gao et al., *Connectivity and Edge Computing in IoT: Customized Designs and AI-based Solutions*, Wireless Networks,
https://doi.org/10.1007/978-3-030-88743-8_2

Industrial communication networks will play a crucial role in the upcoming IIoT [4]. As in the general IoT, machine-type communication (MTC) is a primary enabler of IIoT networks. Features of MTC that are recognized by the 3rd Generation Partnership Project (3GPP) include the following [5]:

- Small packet transmission: MTC devices usually transmit and receive small amounts of data, e.g., 1 kilobyte (KB) data size;
- Time controlled access: MTC devices may tolerate communicating in predefined time intervals to avoid signaling overhead;
- Low mobility: MTC devices either remain at the same location or move infrequently within a limited area;
- Monitoring: the network should be able to detect events such as the loss of connectivity, change of location, and communication failure;
- Group-based features: the network should support the grouping of MTC devices and the association of devices to groups.

Besides the above features mentioned by 3GPP, additional features of MTC have been identified in the literature [6]:

- Uplink-dominated transmissions: The uplink traffic can be caused by a vast number of sensors sending data to an AP;
- Low data rate: Typical data rate for MTC ranges from 100 kilobits per second (kbps) to 10 megabits per second (mbps);
- Sporadic transmissions: Packet inter-arrival time at each device may range from several milliseconds to several minutes [7];
- Low-complexity devices: MTC devices are usually cost-constrained and may not support complex on-board processing.

Compared with the general IoT, IIoT has some unique characteristics. First, connectivity in IIoT is usually structured, featuring centralized network management [8]. Second, IIoT scenarios generally involve densely deployed devices in a relatively limited area. For example, process monitoring in IIoT may involve $10,000$ devices per km^2 [9]. Third, certain IIoT applications are mission-critical and have extremely stringent quality of service (QoS) requirements. For example, the communication latency tolerance for machine tool motion control can be less than 0.5 ms [10]. The combination of the above characteristics poses a significant challenge for supporting MTC in IIoT. Specifically, within a limited geographical area, such as a factory, a communication network may need to support a massive number of devices and, simultaneously, satisfy exceptionally strict QoS requirements for some devices.

Existing standards, including LTE-M, NB-IoT, IEEE 802.11ah, and 5G new radio (NR), are not sufficient for supporting MTC in IIoT networks. For example, LTE-M and NB-IoT, both targeting low-power wide-area communications, are more concerned with radio spectrum usage and power consumption than communication delay. As for 5G NR, the delay threshold to support 10^6 connections per km^2 is set to 10 s in the link-level simulations conducted by 3GPP, while a low packet arrival rate, i.e., 1 packet per 2 h per device, is used [11]. More details on these standards will be given in Sect. 2.2.2. To address the challenge in simultaneously supporting high device density and satisfying stringent QoS requirements, various solutions have been proposed for different layers in the network protocol stack. At the physical layer, utilizing spectrum resources beyond 30GHz or adopting nonorthogonal multiple access (NOMA) can provide support for a high device density. However, physical layer solutions have limitations in terms of transceiver hardware and signal processing complexities, cost-effectiveness, and signaling overhead. At the link layer, new MAC designs and enhancements have been proposed for cellular networks and wireless local area networks, which we will discuss in detail in Sect. 2.2.4.

While the existing studies provide important insights, further research on customized MAC protocols is necessary for smart factories and, in particular, for applications such as factory automation and process control. In the rest of this chapter, we present the communication requirements of the smart factory use case, review the related solutions, tailor a MAC design for smart factories, and demonstrate the performance of our design.

2.2 Connectivity Requirements of Smart Factory

In a smart factory, various devices are connected in order to collect, share, and analyze data for improving the productivity and safety of manufacturing while reducing costs. The smart factory use case involves many applications, most of which require a fast and reliable communication network. In this section, we introduce application-specific requirements, review existing standards, and summarize recent research efforts on enabling smart factories.

2.2.1 Application-Specific Requirements

Different applications have different connectivity requirements. Table 2.1 shows some representative applications and their requirements [10]. In the table, "cycle time" refers to the transmission interval in periodic communication, which is usually larger than the acceptable communication delay. In addition to the applications in this table, more information on application-specific connectivity requirements can be found in Annex E of 3GPP TR 22.804 [12].

Table 2.1 shows that the smart factory use case features a wide range of applications and encompasses a variety of devices such as mobile robots, milling machines, and automated guided vehicles. Accordingly, an IIoT network will need to simultaneously support different applications and their performance requirements. For example, cooperative motion control for mobile robots has a cycle time of 1 ms and thus a stringent communication delay tolerance, while the number of mobile robots in a network is limited. By contrast, process automation or monitoring

Table 2.1 Representative applications and connectivity requirements [10]

Applications		Cycle time	Typical area	Number of devices
Motion Control	Printing machine	<2 ms	100 m × 100 m × 30 m	>100
	Machine tool	<0.5 ms	15 m × 15 m × 3 m	~20
	Packaging machine	<1 ms	10 m × 5 m × 3 m	~50
Mobile robots	Cooperative motion control	1 ms	<1 km^2	100
	Video-operated remote control	10–100 ms	<1 km^2	100
Mobile control panels with safety functions	Assembly robots/ milling machines	4–8 ms	10 m × 10 m	4
	Mobile cranes	12 ms	40 m × 60 m	2
Process automation/monitoring		>50 ms	10^4 devices per km^2	

has a cycle time of 50 ms or more and thus a relatively less stringent communication delay tolerance, while the number of devices to be connected is large. Therefore, an industrial network must achieve the following targets:

- Connecting a large number of devices with assorted types;
- Satisfying different performance requirements, some of which can be highly stringent, for different devices and applications.

From the above two targets, we can see that future industrial networks need to simultaneously support URLLC and mMTC. Next, we review existing standards and recent research efforts related to IIoT.

2.2.2   Related Standards

In recent years, new standards for supporting MTC have been emerging. Existing standards for MTC can be categorized into two groups: cellular-based and non-cellular-based standards. The most representative examples of the former are NB-IoT, LTE-M, and 5G NR, while the most representative examples of the latter are IEEE 802.11ah and proprietary standards such as LoRa. Among these protocols, IEEE 802.11ah is a wireless local area network (WLAN) standard, and the others are wide-area network standards.

NB-IoT debuted in 3GPP Release 13 and can support up to 10^6 connections per km^2 [13] or 5×10^4 connections per cell [14] with a low data rate (160 kbps or less) using a 180 kHz channel bandwidth. Moreover, NB-IoT devices can save device battery power by remaining in low-power mode in-between transmissions. However, the cost for energy efficiency is a large delay. The typical delay of NB-IoT is larger than 1 second (s) and sometimes as large as 10 s, which is unacceptable for many IIoT applications. Compared with NB-IoT, the delay performance of LTE-

M is better but still far from meeting the expectation of IIoT. The LTE-M uses a much larger bandwidth, i.e., 5 MHz as opposed to 180 kHz, and supports a much higher data rate, up to 7 Mbps. The reported connection density of LTE-M varies in the range from 10^4 [14] to 8.5×10^4 connections per cell [15].[1] While the delay performance of LTE-M is better than that of NB-IoT, it is still larger than 100 ms under low network load and can easily increase to 1 s or more under high network load [16].

5G NR can support either massive connections or millisecond-level delay. However, 3GPP Release 15 lists "Critical Communications (CC) and URLLC" and "massive Internet of Things (mIoT)" as two separate service aspects [17]. Consequently, link-level simulations conducted by 3GPP demonstrate that 5G NR can satisfy the URLLC performance requirement of 1 ms delay or the mMTC performance requirement of 10^6 connections per km^2, but not at the same time. Specifically, the delay threshold for NR to support 10^6 connections per km^2 is set to 10 s in the above evaluation, while the packet arrival rate is 1 packet per 2 h per device [11], which is much lower than the cycle times in Table 2.1. Note that 3GPP Release 16 improves support for IoT through enhancements in scheduling and network reference time synchronization. Nevertheless, more improvement is necessary for simultaneously achieving high density and low latency [18].

Using a bandwidth from 125 kHz to 500 kHz, the propriety standard LoRa could achieve a theoretical transmission range of 10 km. However, the probability of successful transmission at the first attempt is below 0.2 when 1000 devices are connected [19]. Such reliability performance can be unacceptable for many smart factory applications. Moreover, the delay of LoRa is no less than 1 s if 100 or more devices are connected and can easily surpass 10 s with further increased connection density [20].[2] The high delay and low reliability limit the use of LoRa in the smart factory use case.

IEEE 802.11ah can support a less than 10 ms delay but only under a low load condition [21]. For example, 802.11ah is suitable for infrequent transmission with a packet inter-arrival duration of no less than 30 s. When 500 devices are connected to the AP, the delay becomes approximately 300 ms even under the relatively long packet inter-arrival time of 180 s [22]. Recent works propose improvements of IEEE 802.11ah, yet the improvement is mostly seen in the throughput rather than in the delay [23, 24]. The lack of guarantee on a millisecond-level delay limits the suitability of IEEE 802.11ah for smart factories.

In summary, the insufficiency of existing standards for supporting MTC in dense industrial networks is clear. None of the existing standards can simultaneously support massive connections and guarantee a millisecond-level access delay, which

---

[1] The upper limit is calculated based on data in [15] as follows: 357,000 devices per 1.08 MHz multiplied by 5 MHz and then divided by 21 cells.

[2] In practice, connection density can be higher than device density due to dual-connectivity or multi-connectivity. We use "connection density" and "device density" interchangeably since multi-connectivity is not a focus of this book.

is needed for the smart factory use case. Therefore, there is an urgent need for developing and standardizing solutions that can meet the needs of the smart factory use case.

2.2.3 Potential Non-Link-Layer Solutions

To address the challenge of supporting high device density and stringent QoS requirements, various solutions have been proposed.

In cellular networks, network densification and network slicing allow networks to support high user density and satisfy stringent QoS requirements. In addition to supporting high connection density, network densification can reduce link access delay [25]. However, there is a limitation on the density of APs due to the increasing cost, and interference and the diminishing performance gain as the network densifies [26]. Network slicing enables flexible service provision for coexisting services with different QoS requirements [27]. As a result, it can potentially contribute to simultaneously supporting mMTC and URLLC in IIoT scenarios. However, the complexity for network slicing can be high due to the need for frequent resource reservation and orchestration [28]. For industrial networks, especially medium or small industrial networks, network densificatoin or network slicing may not be an ideal solution.

Emerging physical-layer techniques, such as millimeter-wave (mmWave) communication and NOMA, could also contribute to supporting the smart factory use case. Extending the available radio spectrum to the mmWave range could help support high connection density. However, equipping every sensor, actuator, and other devices with hardware for mmWave communications can yield a high cost, which may hinder the practical deployment of smart factories. Some research works have proposed NOMA solutions for mMTC, such as compressed sensing based multi-user detection [6], coded tandem spreading [29], and block sparsity and block precoding [30]. Such solutions usually require either advanced signal processing, which increases algorithm and hardware complexity, or availability of transmitter side information at receivers, which results in signaling overhead [6, 31].

2.2.4 Link-Layer Solutions: Recent Research Efforts

Different from physical-layer solutions, link-layer solutions can be flexible as they can be implemented through software. In addition, link-layer solutions can be cost-effective and customized based on the application QoS requirements. As a result, link-layer solutions have tremendous potential for supporting the smart factory use case. Existing studies on link-layer solutions can be categorized into solutions for cellular networks, solutions for WLANs, and hybrid solutions.

For cellular networks, the bottleneck is the contention-based random access channel (RACH) procedure for connection setup. Specifically, a network can be congested when a massive number of devices try to establish connections around the same time [32]. In 3GPP release 15, the design of early data transmission (EDT) replaces a standard four-step RACH procedure with a two-step procedure [17]. In existing research works, prioritization and grouping have been a focus of refining the RACH procedure, and different ideas have appeared. Devices can be grouped, e.g., based on their delay requirements, to limit the collision probability. Groups contend with each other to make access attempts, while either one [33] or multiple groups [34] can be active at a given instant. Alternatively, devices can be grouped and later redistributed into different groups after encountering an initial collision [35]. Access class barring [36], extended access class barring [37], and their derivatives [38–40], as distributed coordinate mechanisms, have also gained popularity and attracted much attention in the literature. After the connection setup stage, scalable transmission time intervals (TTI) [41, 42], and preemption of scheduled low-priority transmissions [43] for high-priority devices can be used for reducing access delay. Overall, most solutions represent refinements of existing protocol designs, e.g., refining the RACH procedure. These solutions often make a trade-off between different performance metrics, e.g., delay and collision probability, while simultaneously improving performance over several metrics is necessary for the smart factory use case. Moreover, the above solutions are grant-based, which causes unnecessary overhead and delay, while grant-free access is preferred [44].

For WLANs, improvements over 802.11ah are the focus of many works on MAC design. The key mechanism in 802.11ah is the restricted access window (RAW), which groups devices and allows channel access for different groups in different time durations. Existing MAC solutions for LAN focus on refining RAW. For example, one category of works optimizes the RAW window size based on the group size [45, 46], dynamically changing the window size according to failed transmission attempts [47], or allocating RAW slots based on group QoS requirements [48]. Another category of works improves the grouping method for RAW according to geographical device distribution [49], traffic volume [50], data rate [23], or potential hidden terminal relationship among nodes [51]. Besides the above two categories, other efforts to improve RAW include periodical RAW for periodic traffic scheduling [52]. Compared with cellular-based solutions, WLAN-based solutions have the advantage of grant-free access and lower overhead, while the disadvantage is lower reliability due to inevitable collisions in data transmission. Consequently, the connection density that can be supported by WLAN-based solutions is generally smaller than that can be supported by cellular-based solutions, while the collision probability and delay can be high in the case of a large number of devices [53].

Moreover, there are hybrid solutions that combine or switch between grant-based access in cellular and grant-free contention-based random access in WLAN. For example, a MAC solution that splits data traffic volume between the two radio access technologies is proposed in [54] for heterogeneous networks in which both cellular and WLAN APs are available. Another example is switching between time-division multiple access (TDMA) and distributed coordination function (DCF), depending on the data traffic load [55]. Additionally, hybrid MAC that implements carrier-sense multiple access (CSMA) and TDMA in different stages of data transmission exists in the literature [56]. These solutions, however, are not customized for smart factories.

2.3   Protocol Design for Smart Factory

In this section, we introduce our protocol design for the smart factory use case in IIoT [57]. We begin with introducing the considered network scenario and then present the elements of our protocol design one by one.

2.3.1   Networking Scenario

Consider a fully connected network with one AP covering a limited geographical area, e.g., a manufacturing facility.[3] A large number of devices such as sensors, actuators, and controllers are densely deployed in the area. The devices are categorized into three types, i.e., high-priority (HP) devices, regular-priority (RP) devices, and low-priority (LP) devices. An illustration of the considered scenario is given in Fig. 2.1.

The overall number of devices and the set of devices are denoted by D and $\mathcal{D}$, respectively. The number and set of HP, RP, and LP devices are denoted by D^H and $\mathcal{D}^H$, D^R and $\mathcal{D}^R$, and D^L and $\mathcal{D}^L$, respectively. Without loss of generality, we assume that the devices are indexed such that the first till the D^Hth devices are the HP devices, the next D^R devices are the RP devices, and the last D^L devices are the LP devices.

Communication Characteristics. The communication characteristics include:

- Short data packets—The length of physical-layer packets is normally in the range between several bytes to several hundred bytes [58];
- Uplink-dominated transmission—A significant portion of the data traffic is attributed to sensor readings or device status reports [6].

QoS Requirements The considered QoS metrics are delay, from the instant of packet arrival to the instant of successful packet transmission, and packet transmission collision probability. Different types of devices have different QoS requirements. Specifically, the maximum tolerable delay and packet collision probability for HP, RP, and LP devices are denoted by δ^H and ρ^H, δ^R and ρ^R, and δ^L and ρ^L

---

[3] The target area is assumed to be less than 1 km^2.

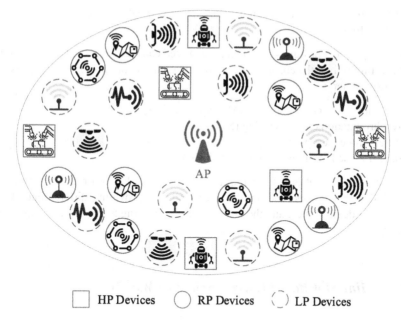

Fig. 2.1 An illustration of the networking scenario

respectively, where $\delta^H < \delta^R < \delta^L$ and $\rho^H < \rho^R < \rho^L$. The value of δ^H is assumed to be small such as on the millisecond level.

Device Packet Arrivals For practicality, we do not assume a specific traffic model. However, we consider the following data packet arrival properties:

- The packet arrival statistics at each device are constant during a relatively long period with respect to packet inter-arrival time. The packet arrival rate of device i in the considered time duration is denoted by λ_i;
- The packet arrival rate is relatively low so that $1/\lambda_i$ is much larger than δ^H for any i. This is in accordance with the sporadic transmission characteristic of MTC, where the packet inter-arrival time can range from tens of milliseconds to several minutes [7];
- For tractability, we assume that the transmission time for data packets is identical and equal to T_x.

Given the networking scenario, we aim to develop a MAC solution with the following features:

(1) Accommodating a large number of devices on a single channel with a single AP;
(2) Satisfying the differentiated QoS requirements for each type of devices;
(3) Keeping control overhead as low as possible;

(4) Exploring the role of machine learning, specifically in device transmission scheduling.

Our MAC protocol design is based on time-slotted channel access, which suits short packets. Tailored for the considered networking scenario, our protocol comprises the following elements:

- Mini-slot based carrier sensing (MsCS);
- Synchronization carrier sensing (SyncCS);
- Differentiated assignment cycles;
- Superimposed mini-slot assignment (SMsA).

The first two elements target at improving channel utilization efficiency through implicit distributed coordination, the third targets at providing differentiated QoS for different device types, and the last targets at increasing the number of supported devices.

2.3.2 Mini-Slot Based Carrier Sensing (MsCS)

Time is partitioned into frames, and each frame is partitioned into n_s slots, as shown in Fig. 2.2. A slot begins with n_m mini-slots, each of length T_m, followed by a duration of length T_x. Accordingly, the length of a slot, denoted by T_s, depends on the number of mini-slots and is equal to $n_m \times T_m + T_x$.

Given the high device density and sporadic transmission pattern, each slot is assigned to multiple devices, in order to achieve high channel utilization efficiency via reducing idle slots. Different devices associated with a slot are assigned different mini-slots of the slot. Different from existing designs with mini-slots, where mini-slots are used for transmitting packets [59, 60] or jamming signals [61], the mini-slots in our protocol are very short, e.g., less than 10 microseconds (μs) and are used for channel sensing instead of sending reservation requests or data packets. In the proposed protocol, the minimum time unit for transmitting a packet is a slot, and each slot accommodates at most one successful packet transmission. Clearly, without proper coordination, transmission collisions may happen when multiple devices are assigned to the same slot.

The purpose of using mini-slots is to enable channel sensing for collision-free distributed channel access. When the AP assigns a slot to a device, it also specifies a mini-slot for the device. Suppose that device i is assigned mini-slot m of slot l. Then, the following rules are used in the proposed protocol:

- If device i has a packet to transmit and $m = 1$, it starts transmitting right away when slot l begins;
- If device i has a packet to transmit and $m > 1$, it needs to sense the channel during mini-slot $m - 1$ of slot l and starts transmitting from mini-slot m of slot l only if the channel is sensed idle. Otherwise, it will skip this slot and wait for the next transmission opportunity;

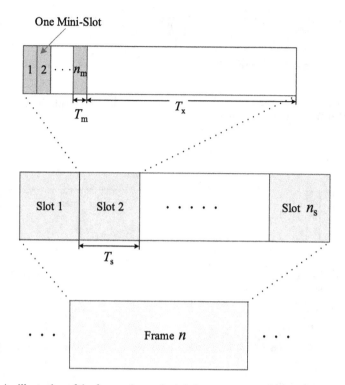

Fig. 2.2 An illustration of the frame, slot, and mini-slot structure

- If device i does not have a packet to send, it simply stays idle in the corresponding slot.

The first two cases are illustrated in Fig. 2.3.

With MsCS, different mini-slots correspond to different transmission priorities. Specifically, a mini-slot with a larger index corresponds to a lower transmission priority. Therefore, mini-slots with small indexes can be used to accommodate HP devices. Via MsCS, a device makes sure that none of the devices with higher priority is using the channel before accessing the channel. As a result, the devices can avoid packet collision while sharing the same slot. Note that the MsCS is fully distributed and does not require any control message exchange, given the assignment of slots and mini-slots to devices by the AP. The cost for avoiding collision is the overhead of using mini-slots for sensing. Specifically, the ratio of usable packet transmission duration over slot length is T_x/T_s.

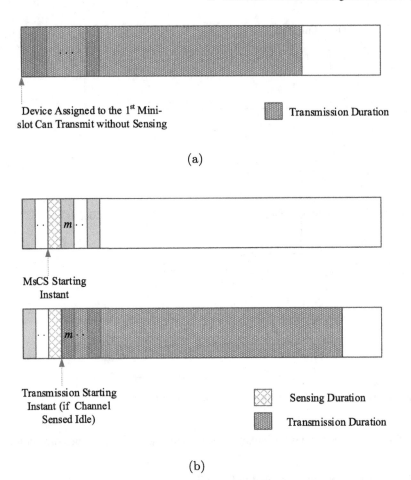

Fig. 2.3 An illustration of the MsCS: (a) Devices assigned to the first mini-slot of any slot starts transmission immediately when the slot begins, without sensing the channel; (b) devices assigned to mini-slot $m(> 1)$ must sense the channel during the $(m - 1)$th mini-slot, and starts transmission at the beginning of the mth mini-slot if the channel is sensed to be idle

For MsCS to work, the following conditions should be satisfied:

- The mini-slot length, T_m, must be longer than the maximum propagation delay across the network coverage area;[4]

---

[4] A possible choice for mini-slot length is $9\,\mu s$, which follows from the DCF slot time in IEEE 802.11ac.

- The overall length of all mini-slots, i.e., $n_m T_m$, should be less than the packet transmission duration T_x. This is for ensuring that each slot accommodates at most one transmission;[5]
- The aggregated packet arrival rate of all devices assigned the same slot must be less than 1 per frame.

2.3.3 Synchronization Sensing (SyncCS)

Even though MsCS improves channel utilization efficiency, as a result of multiple devices sharing each slot, none of the devices may have a packet to transmit in a slot. Increasing the number of mini-slots in each slot can reduce the slot idle probability. However, it may violate the delay requirements for devices assigned high-index mini-slots or the aforementioned condition that $n_m T_m \leq T_x$.

Alternatively, if idle slots can be identified and avoided, the channel utilization efficiency can be further improved, and so will the resulting QoS. To achieve this, the following rules of SyncCS are used in the proposed protocol:

- All devices in $\mathcal{D}$ sense the channel in the last mini-slot, i.e., mini-slot n_m, of everyone slot. The exceptions are: (i) any device that is transmitting and (ii) the device that is assigned mini-slot n_m;[6]
- If the last mini-slot is idle, the rest of the current slot is skipped and the next slot starts immediately after this last mini-slot;
- If the last mini-slot is busy, the next slot starts after the current slot ends.

The above rules are illustrated in Fig. 2.4, and the rationale is explained as follows. Given the condition that $n_m T_m < T_x$ as mentioned in Sect. 2.3.1, no device is or will be transmitting in a slot if the last mini-slot of that slot is idle. Therefore, upon sensing an idle last mini-slot, all devices know that the rest of the slot can be skipped and the next slot can start after this mini-slot. The SyncCS allows devices to synchronize slots even though the length of a slot is no longer fixed. With SyncCS, a busy slot has the full length of $n_m \times T_m + T_x$, while an idle slot has the reduced length of $n_m \times T_m$.

The SyncCS has two main differences from the MsCS:

- In SyncCS, devices must perform sensing regardless of whether they have a packet to transmit or not (with exceptions as mentioned above);
- In SyncCS, all devices, not just the devices assigned to the slot, need to sense the channel in each slot.

---

[5] In an extreme case when a device assigned a low-index mini-slot transmits a very short packet, it is possible that a device assigned a high-index mini-slot senses channel idle and transmits a packet in the same slot. This extreme case is ignored in the protocol design and performance analysis.

[6] The device assigned mini-slot n_m knows whether the slot is idle or not from sensing the channel during mini-slot $n_m - 1$ as mandated by the MsCS.

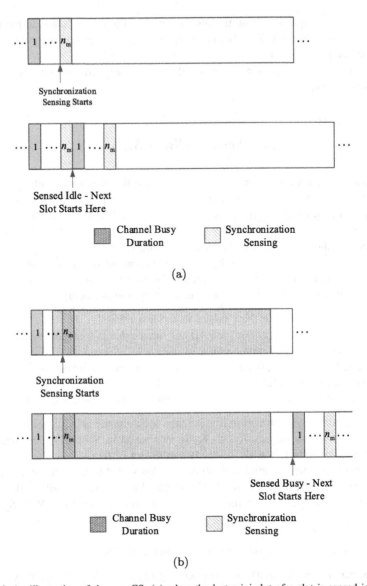

Fig. 2.4 An illustration of the syncCS: (**a**) when the last mini-slot of a slot is sensed idle, the remaining transmission duration of this slot is skipped, and the next slot starts right after the last mini-slot of this slot; (**b**) when the last mini-slot of a slot is sensed busy, the next slot starts after the entire duration of this slot

Similar to the MsCS, SyncCS is fully distributed and does not require any control message exchange. The cost for further improving channel utilization efficiency via SyncCS is the extra channel sensing. In addition, accurate time synchronization is required among all devices. Without SyncCS, a device can be in the sleep mode for most of the time in a frame and only wake up before its assigned mini-slot for MsCS if it has a packet to transmit. With SyncCS, each device needs to perform sensing in each slot and re-synchronize once for each idle slot. In the IIoT scenario under consideration, it is possible that energy consumption of devices is less of a concern (e.g., as compared with sensors deployed in remote areas such as in forests); Otherwise, the design element of SyncCS can be omitted in the proposed protocol.[7]

2.3.4 Differentiated Assignment Cycles

Using the slot structure in Fig. 2.2, the delay for a device depends on the frame length if each device has at most one transmission opportunity in each frame. However, one transmission opportunity in each frame for every device does not provide sufficient flexibility to support differentiated QoS. Particularly, the maximum delay threshold of HP devices, i.e., δ^H, can be much smaller than that of RP/LP devices. To address this problem, we extend the frame in Fig. 2.2 to differentiated assignment cycles. Specifically, each HP, RP, and LP assignment cycle consists of r^H, r^R, and r^L slots, respectively, where $r^H < r^R < r^L$. Each HP, RP, or LP device is assigned one mini-slot of one slot in each HP, RP, or LP assignment cycle, respectively. Thus, an HP/RP/LP cycle serves as a frame for the HP/RP/LP devices, respectively. In the case when all devices have the same priority, the HP, RP, and LP cycles become identical and reduce to a standard frame. The differentiated assignment cycles are illustrated in Fig. 2.5, in which different color patterns in the mini-slots represent different assigned devices. In the illustration, r^L is a multiple of r^R, and r^R is a multiple of r^H.[8] The HP devices assigned to the same slot in any different HP assignment cycles are identical, as shown by the two illustrated slots at the top of Fig. 2.5, while the RP or LP devices assigned to the two slots are different.

With differentiated assignment cycles, it becomes possible to achieve the stringent delay requirement of HP devices, by setting r^H small, and at the same time support a large number of devices, by using a large r^R and/or r^L. Note that similar idea of differentiated cycles can be found in existing works such as [62], where two different cycle lengths are used for realtime and non-realtime traffic, respectively. With a different slot structure and three different cycle lengths, we adopt the same essential idea here. This is because, for scheduling based channel access, achieving lower delay translates to more frequently scheduled transmission opportunities. This naturally leads to differentiated cycles for different device or traffic types.

---

[7] Alternatively, the AP may broadcast frame synchronization beacons. In such case, when a device has a packet to send, it can wake up and synchronize to the next frame. It may remain awake and synchronized to each slot until the packet is transmitted.

[8] While r^L does not have to be a multiple of r^R or r^H in theory, the overall device assignment cycle is the lowest common multiple of r^H, r^R, and r^L. Limiting the lowest common multiple to be r^R itself can reduce the complexity of device assignment by the AP.

2.3.5   Superimposed Mini-slot Assignment (SMsA)

The proposed MAC protocol aims to support a high device density. The MsCS and SyncCS contribute to the solution by improving channel utilization efficiency, along with differentiated assignment cycles with a large r^R and/or r^L. In addition, if devices can share a mini-slot, beyond only sharing a slot, the capacity of the network in terms of the number of supported devices can be significantly improved, at the cost of nonzero packet transmission collision probabilities.

The final element in our proposed protocol, i.e., SMsA, allows the assignment of one mini-slot to multiple devices, provided that packet transmissions associated with such assignment can be properly scheduled as not to violate the QoS requirements of the devices. For the simplicity of presentation, we limit the SMsA to devices of the same type, i.e., an HP device can share a mini-slot only with other HP devices. With SMsA, a mini-slot in Fig. 2.5 may no longer be assigned to a device exclusively.

Transmission collision may happen among devices sharing a mini-slot, and the collision probability depends on the following factors:

- The device packet arrival rates;
- The number of mini-slots and the mini-slot assignment;

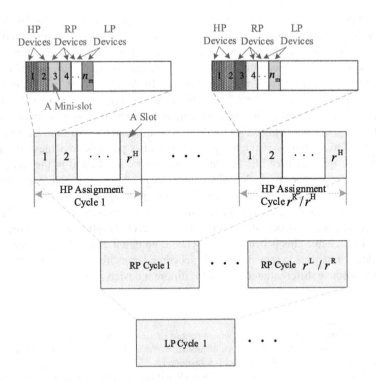

Fig. 2.5 An illustration of differentiated assignment cycles

- The HP, RP, and LP assignment cycle lengths.

While the device packet arrival rates are not controllable, the collision probability may be reduced by properly determining the last two factors (to be studied in Sect. 2.5).

We do not consider collision resolution here. However, a design element for collision detection can be added to our proposed MAC protocol. The following is an example. If two or more devices assigned the same mini-slot simultaneously start sending packets to the AP, the AP will detect the collision. As soon as the AP detects the collision, it will start broadcasting a collision beacon that fills the rest of the current slot. On the device side, the sending devices will switch to sensing mode to check for a collision beacon after transmitting their packets. If a beacon is sensed, the device knows that a packet collision happened during its transmission and may decide to re-transmit the packet in another slot.

2.3.6 Downlink Control

The AP broadcasts the mini-slot and slot assignment to devices via downlink control messages. Based on the assumption of stationary traffic statistics in a relative long duration,[9] the assignment does not need to be updated frequently. The AP may either broadcast the entire assignment in one downlink control message or breakdown the assignment information into multiple messages.

Consider an example of 10 mini-slots per slot (i.e., $n_m = 10$) and 200 slots per LP assignment cycle (i.e., $r^L = 200$). In such case, 2 bytes is more than sufficient to represent the slot and mini-slot assignment for each device. For 1000 devices, the assignment message payload size is no more than 2 kilobytes (kB). For a slot length of 200 µs, an LP assignment cycle is about 40 ms in length. Even if the traffic statistics change as frequently as once in every 5 minutes, the 2 kB downlink assignment message is needed just once in every 7500 LP assignment cycles or, equivalently, 1.5×10^6 slots.

As downlink control messages are infrequent in comparison with the dominating uplink messages, we neglect the impact of downlink control messages while analyzing the performance of the proposed protocol.

The core of MAC protocol design is to coordinate transmissions from devices, while prioritizing and device grouping are two important aspects

(continued)

---

[9] The stationary duration, if denoted by T_{st}, should satisfy $T_{st} \gg r^L T_s$.

of coordination. There are various approaches for prioritizing, such as using both contention-based and contention-free access in a MAC protocol [63]. Similarly, there are many grouping approaches, such as limiting contention to devices generating packets around the same instant [64]. In our proposed MAC protocol, the utilization of mini-slots is inherently capable of both prioritizing and grouping. Meanwhile, the differentiated assignment cycles further strengthen the design's capability in prioritizing, while the SMsA further strengthens its capability in grouping.

2.4 Performance Analysis

In this section, we present performance analysis of the proposed MAC protocol, focusing on the MsCS, SyncCS, and SMsA. Note that the proposed MAC protocol works under the following conditions:

- The expected number of packet arrivals summarized over all devices sharing a slot is less than 1 per frame;[10]
- The average packet arrival interval of any device is larger than the maximum tolerable packet delay of that device.

In practice, some devices can have a high packet arrival rate that violates the above conditions. In such case, more than one slot can be assigned to such a device in the corresponding assignment cycle so that the expected number of packet arrivals of the device per scheduled slot is less than one. In the subsequent analysis, we simply assume that the number of packet arrivals for any device is less than one per its assignment cycle.

Without assuming a specific traffic model, we focus on the first-order statistic. The expected number of packet arrivals at device i in a frame is given by $\lambda_i T_f$, where T_f denotes the length of a frame. Denote the set of all HP, RP, and LP devices assigned to slot l by $\mathcal{D}_l$. Denote the delay of device i, averaged over packet transmissions while the traffic is stationary, by τ_i. The aforementioned two conditions correspond to the following equations:

$$\sum_{i \in \mathcal{D}_l} \lambda_i T_f \leq 1, \forall l, \tag{2.1a}$$

$$\frac{1}{\lambda_i} \geq \tau_i, \forall i. \tag{2.1b}$$

[10] This condition applies to the case without differentiated assignment cycles. In the case with differentiated assignment cycles, the condition is different.

2.4.1 Delay Performance with No Buffer

We investigate the impact of mini-slots in the case without SMsA, given the slot assignment and device packet transmission probabilities (estimated from the packet arrival rates). Starting from a simplified scenario, the analysis here is based on the following assumptions:

- The condition in (2.1a) is satisfied;
- A packet not in transmission is dropped when a new packet is generated. The scenario where devices have buffers is analyzed in Sect. 2.4.2;
- All devices are of the same type and priority. Consequently, the three assignment cycles reduce to a unified frame with n_s slots;
- The SyncCS is not adopted. The analysis of SyncCS is given in Sect. 2.4.4.

We focus on the delay analysis since collision probability is zero without SMsA. Let τ_0 denote the *base delay*, defined as the time duration from the packet arrival instant till the first assigned mini-slot. Under the aforementioned assumptions, the average base delay is equal to $n_s T_s / 2$ for all devices, as each device has one assigned mini-slot in each frame. The *overall delay* is the base delay plus the *access delay* (AD), i.e., the duration from the first assigned mini-slot since the packet arrival till the end of the packet transmission. Since the average base delay is a constant here, we focus on finding the average AD.

Denote the device assigned the mth mini-slot of the lth slot by $d_{m,l}$. Denote by $\tau_{m,l}$ the average *access delay counted in frames* (AD-F), i.e., the number of logical frames since device $d_{m,l}$'s packet arrival till device $d_{m,l}$'s packet transmission.[11] Different from a physical frame, a logical frame counted in the AD-F for device $d_{m,l}$ is the duration from slot l of one physical frame to slot l of the next physical frame. Therefore, a logical frame has the same length as a physical frame, but different starting and ending points for different devices. Accordingly, the arrival and transmission of a packet can happen within one logical frame, and the resulting AD-F is 1 in such case.[12] Note that AD-F $\tau_{m,l}$ corresponds to a duration slightly longer than the AD defined in the preceding paragraph. This is because the AD ends when a packet transmission is completed, while the AD-F counts the entire frame into the delay, including the duration after device $d_{m,l}$'s packet transmission. Accordingly, the AD of device $d_{m,l}$ can be obtained from the AD-F by calculating $(\tau_{m,l} - 1) \times T_f + T_x$, where the frame length T_f is equal to $n_s T_s$.

---

[11] When packet re-transmission is considered, the definition of AD-F should be changed to "the number of logical frames since packet arrival till successful packet transmission". Meanwhile, the "packet arrival rate" in our analysis should be replaced by "packet transmission rate" as a packet may need re-transmission(s).

[12] In the rest of this chapter, we do not distinguish physical and logical frames and refer to both as "frame" since they are equal in length.

Since any device assigned the first mini-slot of any slot can transmit right away without sensing when the slot begins, we have

$$\tau_{1,l} = 1, \forall l. \tag{2.2}$$

For devices assigned the subsequent mini-slots, the AD-F can be found using the following result.

Theorem 2.1 *For any integer m such that $1 \leq m \leq n_{\mathrm{m}} - 1$, the following relation between the AD-F of device $d_{m+1,l}$ and device $d_{m,l}$ holds:*

$$\tau_{m+1,l} = \frac{1}{1 - \gamma_{m,l} - T_{\mathrm{f}}\lambda'_{m,l}} \left(-\frac{(1 - \gamma_{m,l})T_{\mathrm{f}}\lambda'_{m,l}}{2}\tau^2_{m,l} \right.$$
$$\left. +\left(1 - \gamma_{m,l} + T_{\mathrm{f}}\lambda'_{m,l}\right)\tau_{m,l} - \frac{T_{\mathrm{f}}\lambda'_{m,l}(1+\gamma_{m,l})}{2} \right) \tag{2.3}$$

where

$$\lambda'_{m,l} = \frac{\lambda_{m,l}}{\left(1 + T_{\mathrm{f}}\lambda_{m,l}(\tau_{m,l} - 1/2)\right)} \tag{2.4}$$

represents the effective packet arrival rate of device $d_{m,l}$ excluding dropped packets due to packet replacement (as there is no buffer), and

$$\gamma_{m,l} = T_{\mathrm{f}} \sum_{r=1}^{m} \lambda'_{r,l} \tag{2.5}$$

represents the expected overall number of packet arrivals in a frame for devices $d_{1,l}$ till $d_{m,l}$ (excluding replaced packets).

The proof of the above theorem can be found in [57]. Using the fact that $\tau_{1,l}$ is equal to 1 for any l, (2.3) can be used to obtain the AD-F for devices assigned to all subsequent mini-slots in a slot recursively.

2.4.2   Delay Performance with Buffer

Now consider the case when each device has a buffer. Recall that different mini-slots correspond to different transmission priorities. In the proposed protocol, any proper slot and mini-slot assignment ensures that the expected number of packets in the buffer of device $d_{m,l}$ is less than one, for any $m < n_{\mathrm{m}}$ and any l. The reason is that, if the expected number of buffered packet at $d_{m,l}$ is larger than or equal to one, devices assigned mini-slots $m + 1, \ldots, n_{\mathrm{m}}$ of slot l have almost no opportunity to transmit. As a result, we neglect the case when there are more than one packet in a buffer and use the following approximation. Specifically, at any instant, a device is in one of three states:

- no packet;
- one packet, transmitting or waiting for channel access;
- two packets, one transmitting or waiting for channel access and the other arriving and going into the buffer.

Accordingly, for any given device, there is either no packet or one packet transmitting or waiting for channel access when a new packet arrives.

Denote by $\tau_{m,l}^{b}$ the average AD-F of device $d_{m,l}$ in the case with buffer, the following result is in order.

Theorem 2.2 *In the case with buffers, for any integer m such that $1 \leq m \leq n_{\mathrm{m}} - 1$, the relation between the AD-F of device $d_{m+1,l}$ and device $d_{m,l}$ is given by*

$$
\tau_{m+1,l}^{b} = \frac{1 - \gamma_{m,l}^{b}}{1 - \gamma_{m+1,l}^{b}} \left(\frac{1}{1 - \gamma_{m,l}^{b} - T_{\mathrm{f}}\lambda_{m,l}} \left(-\frac{(1 - \gamma_{m,l}^{b})T_{\mathrm{f}}\lambda_{m,l}}{2} \right. \right.
$$

$$
\left. \left. \cdot \left(\tau_{m,l}^{b} \right)^{2} + (1 - \gamma_{m,l}^{b} + T_{\mathrm{f}}\lambda_{m,l})\tau_{m,l}^{b} - \frac{T_{\mathrm{f}}\lambda_{m,l}(1 + \gamma_{m,l}^{b})}{2} \right) - 1 \right) + 1
$$

where

$$
\gamma_{m,l}^{b} = T_{\mathrm{f}} \sum_{r=1}^{m} \lambda_{r,l} \tag{2.6}
$$

represents the expected overall number of packet arrivals in a frame for devices $d_{1,l}$ till $d_{m,l}$.

The proof of the above theorem can be found in [57].

2.4.3 Slot Idle Probability

A slot is idle if none of its associated devices transmits. Under stationary packet arrival statistics, the expected slot idle probability of MsCS can be obtained. In the case with and without buffer, the slot idle probability is approximately given by

$$
\eta_{l}^{b} = 1 - \sum_{m=1}^{n_{\mathrm{m}}} \lambda_{m,l}T_{\mathrm{f}} \tag{2.7a}
$$

$$\eta_l = 1 - \sum_{m=1}^{n_m} \lambda'_{m,l} T_f \qquad (2.7b)$$

respectively, where $\lambda'_{m,l}$ is given in (2.4). Note that the right-hand side of either of the two equations above is non-negative when the condition (2.1a) is satisfied, i.e., when the slot is not overloaded. The above approximation of slot idle probability also assumes a negligible packet collision probability, i.e., the expected number of transmitted packets and the expected number of packet arrivals (that cause no packet replacement) are equal in any slot.

Define the throughput of slot l as the expected number of packets transmitted in the slot. The slot throughput equals $1 - \eta_l^b$ and $1 - \eta_l$ for the cases with and without buffers, respectively.

2.4.4   Impact of SyncCS

As SyncCS results in two possible lengths for each slot, i.e., the full and the reduced lengths, the frame length becomes a random variable. Denote the expected frame length with SyncCS in the case with and without buffer by $T_f^{e,b}$ and T_f^e, respectively. Denote n_s' as the number of busy slots out of the n_s slots in a frame. In the case without buffer, it follows that

$$T_f^e = n_s n_m T_m + n_s' T_x. \qquad (2.8)$$

Since there is no collision,

$$T_f^e \sum_l \sum_m \lambda'_{m,l} = n_s' \qquad (2.9)$$

because the expected number of packet transmissions should equal the expected number of arriving packets (that are not replaced) in a frame duration. From (2.8), (2.9), and (2.4) (with T_f replaced by T_f^e), n_s' and T_f^e can be solved.

In the case with buffer, we have

$$T_f^{e,b} = n_s n_m T_m + n_s' T_x \qquad (2.10a)$$

$$T_f^{e,b} \sum_l \sum_m \lambda_{m,l} = n_s' \qquad (2.10b)$$

which gives

$$T_f^{e,b} = \frac{n_s n_m T_m}{1 - \sum_l \sum_m \lambda_{m,l} T_x}. \qquad (2.11)$$

Substituting T_f in (2.3) and (2.6) with T_f^e and $T_f^{e,b}$, respectively, gives the AD-F of the proposed design with MsCS and SyncCS. In the case without buffer, T_f^e depends on $\tau_{m,l}$ through (2.4), which renders a complicated relation.

2.4.5 Impact of SMsA

The AD-F in Sects. 2.4.1 and 2.4.2 is obtained when each mini-slot is assigned to a device exclusively. With SMsA, we have the following questions:

- What is the relation among the AD-F of different devices assigned the same mini-slot?
- How does the SMsA impact the relation in the AD-F between devices assigned adjacent mini-slots?

Denote the set of all devices assigned mini-slot m of slot l by $\mathcal{D}_{m,l}$. The following theorem answers the first question.

Theorem 2.3 *In the case without buffer, all devices in $\mathcal{D}_{m,l}$ have the same AD-F, regardless of the difference in their individual packet arrival rates. In the case with buffer, assuming a negligible packet collision probability and*

$$\lambda_i \ll \sum_{r=1}^{m} \sum_{j \in \mathcal{D}_{r,l}} \lambda_j, \forall i \in \mathcal{D}_{m,l}, \tag{2.12}$$

the differences among the AD-Fs of devices in $\mathcal{D}_{m,l}$ are negligible.

For the second question, similar to (2.3) and (2.6), the relation between the AD-Fs of devices in adjacent mini-slots in the case of SMsA can be characterized. The characterization is given in [57]. It is worth mentioning that the packet collision probability has an impact on the AD-F even if devices do not detect collisions or re-transmit. Given the aggregated packet arrival rate of devices sharing a mini-slot, a higher collision probability implies less channel busy duration for transmitting the same amount of packets. Consequently, the average packet waiting time and the AD-F decrease as the collision probability increases. However, if the collision probability is low, such impact can be negligible.

With the AD-F, we can estimate the packet collision probability. Consider the case with buffer as an example and assume that the condition in (2.12) is satisfied. Based on Theorem 2.3, all the devices in $\mathcal{D}_{m,l}$ have the same AD-F, denoted by $\tau_{m,l}^{s,b}$. Then, any device in $\mathcal{D}_{m,l}$ with a packet to send is expected to have one transmission opportunity in every $\tau_{m,l}^{s,b}$ frames. The expected number of packet arrivals at device $i \in \mathcal{D}_{m,l}$ between any two consecutive transmission opportunities, which must be less than 1, can be estimated by $\tau_{m,l}^{s,b} T_f \lambda_i$. With the MsCS, all devices in $\mathcal{D}_{m,l}$ that have packets to send share the same transmission opportunities. Therefore, the probability that device i's packet encounters a collision is approximately given by

$$q_i^{c,b} = 1 - \prod_{j \in \mathcal{D}_{m,l} \setminus \{i\}} \left(1 - \tau_{m,l}^{s,b} T_f \lambda_i\right). \tag{2.13}$$

Note that, knowing only the average packet arrival rates, the above approximation may be limited in accuracy. An accurate determination of the collision probability requires the traffic arrival model of all devices, which can be difficult to obtain in practice. In Sect. 2.6, we will demonstrate through numerical results that our approximation can be a useful tool for device assignment.

2.5 Scheduling and AI-Assisted Protocol Parameter Selection

While the proposed MAC design gives the frame of our connectivity solution for the smart factory use case, the performance of our solution also depends on the scheduling of transmission opportunities for the devices. In this section, we cover the background of scheduling, introduce the scheduling problem in our design, and develop AI-assisted protocol parameter selection for the scheduling component in our connectivity solution.

2.5.1 Background

Section 2.3 introduces our MAC protocol for MTC in IIoT, which provides a potential to increase network capacity and improve QoS performance through increasing channel utilization efficiency. Meanwhile, how to utilize this potential to *guarantee* stringent QoS requirements in a dense network calls for further investigation. Specifically, given the proposed mini-slot based slot structure and a large number of devices, proper *scheduling*, i.e., determining the slot/cycle lengths and assigning the devices specific slots and mini-slots, has a significant impact on the MAC performance.

In our networking scenario, scheduling is for single-hop and uplink communications. Even in this limited scope, many research works exist in the literature, with a common focus on the trade-off between performance and signaling overhead. Early studies include the development of semi-persistent scheduling for voice over IP in Long Term Evolution (LTE) [65], which aims to achieve a balance between system capacity and signaling overhead. For WLAN, Wang and Zhuang propose a token-based scheduling scheme, which achieves performance prioritization for different traffic types with a low overhead in a fully connected network [66]. Gamage et al. develop uplink scheduling for WLAN and cellular interworking to enable multi-homing voice and data services [67].

Despite the abundance of existing studies, scheduling in the setting of MTC and IIoT remains challenging. Ksentini et al. note the potentially overwhelming overhead in the uplink scheduling with a massive number of MTC connections and consider a simple round-robin scheduling algorithm for the case with no QoS requirements [68]. Lioumpas et al. recognize that schedulers designed for general cellular networks cannot be directly applied to MTC, due to a higher device density and a wider variety of QoS requirements, and propose a scheduling algorithm to prioritize devices with low delay tolerance [69]. However, the delay requirements considered therein is in the range from 10 ms to 10 minutes, which can be too large for IIoT applications.

To handle a large number of devices, a popular strategy is to divide the devices into groups (or clusters) and schedule the devices based on the groups [70]. Si et al. propose a grouping-based algorithm that adjusts the service rate for each user group to provide statistical QoS guarantees, where the considered delay requirements are in the range from 20 ms to 100 ms [71]. Karadag et al. present semi-persistent scheduling for MTC in cellular networks, taking delay constraints of devices into account, where devices have periodic traffic arrivals [72]. Zhang et al. propose a random access scheme for MTC in cellular networks by grouping devices according to their delay requirements and applying access control for each group based on the group size, aggregated packet arrival rate, etc. [73]. Arouk et al. propose a group paging based scheduling for massive MTC access in cellular networks, where the key idea is to scatter the contention for channel access to improve performance in terms of delay, collision probability, and energy consumption [74]. The focuses of the last two works are on throughput maximization and energy consumption reduction, respectively.

Given a high device density, diversified service types, and stringent QoS requirements, scheduling may need to be further fine-grained. Specifically, a scheduler may need to attend to the available information (e.g., packet arrival rate) or access strategy of each single device. Salodkar et al. propose a learning-assisted scheduling scheme, in which each device uses reinforcement learning to determine a preferred transmission rate and a base station (BS) schedules the device with the highest rate [75]. Such a scheme can adapt to unknown packet arrival statistics. Chang et al. propose device-level uplink scheduling schemes based on conflict-avoiding codes, in which each device is assigned a two-dimensional code matrix [76]. These schemes are applicable when multiple channels are available. In their recent work, Rodoplu et al. present proactive forecasting-assisted scheduling to support massive access in the IoT, which explores machine learning to predict the traffic of each device and reserve channel time accordingly [77]. The scheme improves network performance with low overhead. Yang et al. utilize a neural network to predict the number of IoT devices and Wi-Fi users, which facilitates dynamic scheduling and channel allocation for co-existing IoT and Wi-Fi communications [78].

In this section, our objective is to develop an effective scheduling scheme to pair with the proposed protocol in Sect. 2.3. Different from the existing works, we focus on achieving QoS guarantee with very low delays. As a part of our MAC protocol, the scheduling scheme contributes to a customized link-layer solution to MTC in

IIoT, supporting high device density, diversified service types, and stringent QoS targets. While we aim to maximize channel utilization efficiency through delicate *distributed* coordination in the MAC protocol in Sect. 2.3, the focus in this section is to develop a *centralized* analysis-based scheduling scheme. The scheduling scheme should achieve a desired balance in the QoS of different services or different QoS metrics for the same service. The integration of distributed coordination and centralized control is expected to strengthen the proposed MAC protocol.

Scheduling for a dense network with hundreds or even thousands of devices can be beyond the reach of conventional approaches, when the packet arrival rate of each device may impact the protocol parameters and the QoS requirement of each device needs to be satisfied. This motivates us to exploit neural networks to assist scheduling. We propose to schedule in two steps, i.e., slot/mini-slot assignment and protocol parameter selection, and develop methods to reduce complexity in each step. The main contribution of this part is twofold: first, we develop algorithms to assign devices specific slots and mini-slots of the proposed protocol in Sect. 2.3, when the protocol parameters are given. Based on the analytical results in Sect. 2.3, the proposed algorithms sort devices of each type, estimate the impact of potential assignments for each device, and make assignments for the devices one by one. As a result, the assignments possess the due accuracy and granularity necessary for satisfying diverse and stringent QoS requirements; Second, to determine the protocol parameters, we exploit a deep neural network (DNN) to assist scheduling. The DNN is structured such that it can be used given any number of devices and learn the mapping from various combinations of device and packet arrival profiles and protocol parameter settings to the resulting scheduling performance. We demonstrate that, after sufficient training, the DNN can learn the mapping. Then, given a specific device and packet arrival profile, the DNN can be used to compare different protocol parameter settings and determine proper parameters for the proposed MAC.

2.5.2 The Considered Scheduling Problem

The following factors have significant impact on the performance of the proposed MAC protocol:

- The number of mini-slots in each slot, i.e., n_m;
- The assignment cycles, r^H, r^R, and r^L, which serve as different frame lengths for different types of devices;
- The device assignment, i.e., the allocation of devices to slots and mini-slots.

We refer to the problem of determining the above factors with the objective of satisfying QoS requirements as the packet transmission scheduling problem, which is illustrated in Fig. 2.6. The AP in the network is expected to have computing capability and conduct the scheduling.

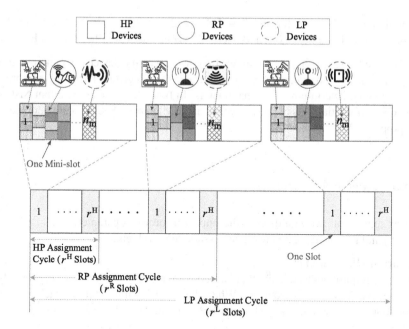

Fig. 2.6 An illustration of the scheduling problem. Different colors in the sub-blocks of a mini-slot correspond to different devices assigned that mini-slot, while dot-filled, solid-filled, and grid-filled patterns represent mini-slots assigned to HP, RP, and LP devices, respectively. The scheduling problem involves determining protocol parameters n_m, r^H, r^R, and r^L as well as assigning slots and mini-slots to all devices

Note that the scheduling problem may not be always feasible. Indeed, we cannot guarantee the satisfaction of arbitrary QoS requirements given an arbitrarily large set of devices with arbitrary packet arrival rates. Thus, the objective here is to investigate effective scheduling that can support as many devices as possible while satisfying their QoS requirements.

Given the sets of all devices $\mathcal{D} = \{1, \ldots, D\}$, HP devices $\mathcal{D}^H$, RP devices $\mathcal{D}^R$, LP devices $\mathcal{D}^L$, and packet arrival rates $\{\lambda_i\}, i \in \mathcal{D}$, we attempt to accommodate all devices while satisfying delay requirements δ^H, δ^R, and δ^L and packet collision probability requirements ρ^H, ρ^R, and ρ^L for the HP, RP, and LP devices, respectively. Based on the protocol, the following constraints exist for the scheduling problem (see Sect. 2.3.4):

- The LP assignment cycle length, r^L, is a multiple of the RP assignment cycle length, r^R, which in turn is a multiple of the HP assignment cycle length, r^H;
- A mini-slot should not accommodate more than one type of devices;
- If mini-slot m of slot l, where $l \leq r^H$ and $m \leq n_m$, is assigned to a subset of HP devices, $\mathcal{I}^H$, then mini-slot m of slot l', for any $l' \in \{r^H + l, 2r^H + l, \ldots, r^L - r^H + l\}$, is also assigned to the same set of HP devices $\mathcal{I}^H$. If mini-slot m of slot l, where $l \leq r^R$ and $m \leq n_m$, is assigned to a subset of RP devices, $\mathcal{I}^R$, then

mini-slot m of slot l', for any $l' \in \{r^R + l, 2r^R + l, \ldots, r^L - r^R + l\}$, is also assigned to the same set of RP devices, $\mathcal{I}^R$. Both cases are illustrated in Fig. 2.6.

To solve the scheduling problem, we first investigate the device assignment while assuming protocol parameters n_m, r^H, r^R, and r^L are given. Then, we explore a DNN to assist determining these parameters. In both steps, we assume that MsCS, SyncCS, differentiated assignment cycles, and SMsA from Sect. 2.3 are adopted in the proposed MAC protocol.

2.5.3   Device Assignment

In this subsection, we first discuss the impact of protocol parameters (n_m, r^H, r^R, and r^L) and then investigate the device assignment problem.

Impact of r^H, r^R, and r^L
The delay requirements, δ^H, δ^R, δ^L, place constraints on r^H, r^R, and r^L, respectively. Consider HP devices for example. When there are n_m mini-slots in each slot, an upper bound on the number of slots per HP assignment cycle, i.e., r^H, is given by[13]

$$\bar{r}^H = \left\lfloor \frac{2\delta^H}{n_m T_m + T_x} \right\rfloor \tag{2.14}$$

where $\lfloor \cdot \rfloor$ is the floor function. The denominator is the length of a slot. The factor '2' in the numerator follows from the fact that the average gap between the beginning of an HP cycle and the arrival of an HP packet is equal to one half of an HP cycle.

Using (2.14), a relation between n_m and r^H can be obtained. If n_m is large, r^H should be small, and the HP devices will be "densely" packed into the r^H slots. As a result, it can be challenging to satisfy the QoS requirements of HP devices. On the other hand, if n_m is small so that r^H can become large, more slots are available for HP devices in each frame. However, the transmission opportunity for RP and LP devices will decrease. Therefore, determining appropriate values for r^H, r^R, and r^L is crucial but nontrivial.

Device Assignment
The assignment of slots and mini-slots to devices is a complex problem. Consider the case with buffer and SMsA. Even if n_m, r^H, r^R, and r^L are given, the device assignment is a combinatorial integer programming problem. Based on the analysis in Sect. 2.4.5, assigning any new device an occupied mini-slot can affect the delay and collision probability of all other devices assigned that mini-slot.

---

[13] The upper bound is obtained under the assumption that every HP device is assigned the first mini-slot of a slot.

We propose a heuristic algorithm for device assignment, built on the analysis in Sect. 2.4, when n_{m}, r^{H}, r^{R}, and r^{L} are given. The analysis allows us to estimate the delay and collision probability of devices in a mini-slot after adding each new device to the mini-slot. The proposed assignment algorithm tentatively assigns a device while estimating the resulting performance, with the target of satisfying the QoS requirements of all assigned devices in the process. The following setting is considered in the assignment:

- All devices assigned the same mini-slot have the same priority type;
- The maximum packet collision probability among all devices assigned the same mini-slot is referred to as the collision probability for that mini-slot and denoted by $q^{\mathrm{c}}_{m,l}$ for mini-slot m of slot l;
- Under the assumption that the impact of collision probability on the cycle length is negligible, the length of an LP cycle can be calculated by

$$T_{\mathrm{f}}^{\mathrm{L}} = \frac{r^{\mathrm{L}} n_{\mathrm{m}} T_{\mathrm{m}}}{1 - \sum\limits_{i \in \mathcal{D}} \lambda_i T_{\mathrm{x}}}, \tag{2.15}$$

which is based on (2.11). Parameter n_{s} in (2.11), i.e., the number of slots in a general frame, is replaced with r^{L} in (2.15) since an LP cycle serves as a frame for LP devices. Note that the use of differentiated assignment cycles does not change the packet arrival rates. Based on the constraints mentioned in Sect. 2.5.2, all devices should be scheduled at least once in an LP cycle, which leads to the summation over the packet arrival rates of all devices in the denominator of (2.15).

Let $\hat{m}_l$ denote the minimum index among the mini-slots of slot l that have not been assigned to any device. For notation simplicity, we omit subscript l in $\hat{m}_l$ when $\hat{m}_l$ and l both appear in the subscript (e.g., $q^{\mathrm{c}}_{\hat{m}_l,l}$ will be written as $q^{\mathrm{c}}_{\hat{m},l}$). The length of the HP, RP, and LP assignment cycles are denoted by $T_{\mathrm{f}}^{\mathrm{H}}$, $T_{\mathrm{f}}^{\mathrm{R}}$, and $T_{\mathrm{f}}^{\mathrm{L}}$, respectively. The proposed assignment is given in Algorithms 1 and 2. Algorithm 1 is the core algorithm for assigning slots and mini-slots to a set of devices with the same priority for a given cycle length, while Algorithm 2 is the overall algorithm that calls Algorithm 1 to make assignments for all devices and all cycles.

In the two algorithms, variables n_i^{c}, $\Lambda_{m,l}$, and $\Gamma_{m,l}$ denote the expected number of simultaneously transmitting packets given that device i is transmitting (which can be larger than 1 as a result of a nonzero collision probability), the aggregated packet arrival rate for all devices assigned mini-slot m of slot l, and the accumulated number of packet arrivals for all devices assigned mini-slots 1 to m of slot l in the corresponding cycle, respectively. Detailed description can be found in Appendix C of [57] and is omitted here for brevity.

The basic ideas of Algorithms 1 and 2 are given as follows. Algorithm 1 assigns mini-slots to devices, starting from the first mini-slot of every slot, and tracks the current mini-slot being assigned. It tentatively assigns a device the current mini-slot

---

Algorithm 1 Core assignment algorithm

Input: $\mathcal{D}^{\dagger}, \mathcal{R}^{\dagger}, n_{\mathrm{m}}, T_{\mathrm{m}}, T_{\mathrm{x}}, \{\lambda_i\}_{\forall i \in \mathcal{D}^{\dagger}}, r^{\dagger}, \hat{m}_l, \forall l, \Gamma_{\hat{m},l}, \forall l.$
Output: Assignment matrix $\mathbf{A}^{\dagger}$ with size $2 \times |\mathcal{D}^{\dagger}|$.
1: *Initialize*: a) $q_{m,l}^{\mathrm{c}} = 0, \forall m, l; n_i^{\mathrm{c}} = 0, \Lambda_{\hat{m},l} = 0, \forall l;$
2:           b) Number of assigned devices $N_{\mathrm{a}}^{\dagger} = 0.$
3:
4: **for** device i in $\mathcal{D}^{\dagger}$ **do**
5:     Check $\tau_{\hat{m},l}, \forall l \in \mathcal{R}^{\dagger}.$
6:     **if** $\min_{l \in \mathcal{R}^{\dagger}} (\tau_{\hat{m},l} - 1) \times T_{\mathrm{f}}^{\dagger} + T_{\mathrm{x}} + \tau_0^{\dagger} > \delta^{\dagger}$ **then**
7:         Quit with flag $F = i;$
8:     **else**
9:         Find set $\mathcal{S}^{\dagger} = \{l|(\tau_{\hat{m},l} - 1) \times T_{\mathrm{f}}^{\dagger} + T_{\mathrm{x}} + \tau_0^{\dagger} \le \delta^{\dagger}\}.$
10:    **end if**
11:    Calculate $\bar{q}_{\hat{m},l}^{\mathrm{c}}$ for tentative assignment $\{\hat{m}_l, l\}, \forall l \in \mathcal{S}^{\dagger}$, using either (2.16a) or (2.17a) with $\tilde{q}_{m,l}^{\mathrm{c}}$ replaced by $\bar{q}_{\hat{m},l}^{\mathrm{c}}$, depending on whether device i is the first device assigned this mini-slot.
12:    **if** $\min_{l \in \mathcal{S}^{\dagger}} \bar{q}_{\hat{m},l}^{\mathrm{c}} > \rho^{\dagger}$ and $\hat{m}_l = n_{\mathrm{m}}, \forall l \in \mathcal{S}^{\dagger}$ **then**
13:        Quit with $N_{\mathrm{a}}^{\dagger} = i;$
14:    **else if** $\min_{l \in \mathcal{S}^{\dagger}} \bar{q}_{\hat{m},l}^{\mathrm{c}} > \rho^{\dagger}$ and $\exists l \in \mathcal{S}^{\dagger} : \hat{m}_l < n_{\mathrm{m}}$ **then**
15:        Update $\mathcal{R}^{\dagger} = \{l \in \mathcal{S}^{\dagger}|\hat{m}_l < n_{\mathrm{m}}\};$
16:        Update $\hat{m}_l = \hat{m}_l + 1$, calculate $\tau_{\hat{m},l}$, and go to Step 3;
17:    **else**
18:        Find slot $l^{\star} = arg \min_{l \in \mathcal{S}^{\dagger}} \bar{q}_{\hat{m},l}^{\mathrm{c}};$
19:        $\mathbf{A}^{\dagger}(1, i) = l^{\star}, \mathbf{A}^{\dagger}(2, i) = \hat{m}_{l^{\star}};$
20:        Update $q_{\hat{m}_{l^{\star}}, l^{\star}}^{\mathrm{c}}$ by setting $q_{\hat{m}_{l^{\star}}, l^{\star}}^{\mathrm{c}} = \bar{q}_{\hat{m}_{l^{\star}}, l^{\star}}^{\mathrm{c}};$
21:        Update $n_i, \Lambda_{\hat{m}_{l^{\star}}, l^{\star}}, \Gamma_{\hat{m}_{l^{\star}}, l^{\star}}$ using (2.16b) to (2.16d) or (2.17b) to (2.17d).
22:    **end if**
23: **end for**
24: **Return** $\{\hat{m}_l\}_{\forall l}, \{\Gamma_{\hat{m},l}\}_{\forall l}, \mathbf{A}^{\dagger}, N_{\mathrm{a}}^{\dagger}.$

---

of all available slots, trying to find the best assignment based on the resulting delay and packet collision probability estimations. If the current mini-slot in none of the slots can accommodate the device in satisfying its collision probability requirement, the algorithm moves to the next mini-slot. The procedure repeats until any of the following three conditions is satisfied: (i) all devices are allocated, (ii) there is no more vacant mini-slot, or (iii) none of current mini-slots can satisfy the delay requirement of a device. Algorithm 2 sorts the devices and calls Algorithm 1 for mini-slot and slot assignment for each device priority type. After obtaining an assignment for HP devices and RP devices, Algorithm 2 extends the assignment for the RP cycle and LP cycle, respectively. Some details of main steps in the algorithms are summarized as follows:

- Step 3 of Algorithm 1—The left-hand side of the inequality represents the overall delay including the base and access delays. The calculation is discussed in Sect. 2.4.1;

Algorithm 2 Overall assignment algorithm

Input: n_m, r^H, r^R, r^L, T_m, T_x, $\mathcal{D}^H$, $\mathcal{D}^R$, $\mathcal{D}^L$, $\{\lambda_i\}_{\forall i \in \mathcal{D}}$.
Output: Device assignment matrix $\mathbf{A}$ (size $2 \times D$), Assignment success flag F_s.
1: *Initialize:* $i = 1$, $q_{m,l}^c = 0$, $\forall m, l$, $n_j^c = 0$, $\forall j \in \mathcal{D}$, $F_s = 0$; Set $\mathbf{A}^R$ and $\mathbf{A}^L$ to all-zero matrices with sizes $2 \times D^R$ and $2 \times D^L$, respectively.
2: Calculate the LP Cycle length using (2.15). Calculate the RP and HP Cycle length using $T_f^R = T_f^L r^R / r^L$ and $T_f^H = T_f^L r^H / r^L$, respectively.
3:
4: Calculate the base delay for HP, RP, and LP devices using $\tau_0^H = T_f^H / 2$, $\tau_0^R = T_f^R / 2$, $\tau_0^L = T_f^L / 2$, respectively.
5:
6: Sort devices in an increasing order of packet arrival rate for $\mathcal{D}^H$, $\mathcal{D}^R$, and $\mathcal{D}^L$, respectively.
7: Set $\hat{m}_l = 1$, $\Gamma_{\hat{m},l} = 0$, and $\tau_{\hat{m},l} = 1$, $\forall l$. Set $\mathcal{D}^\dagger = \mathcal{D}^H$, $\mathcal{R}^\dagger = \{1, 2, \ldots, r^H\}$, $T_f^\dagger = T_f^H$, $r^\dagger = r^H$, $\tau_0^\dagger = \tau_0^H$, $\delta^\dagger = \delta^H$, and $\rho^\dagger = \rho^H$. Run Algorithm 1 and output $\{\hat{m}_l\}_{\forall l}$, $\{\Gamma_{\hat{m},l}\}_{\forall l}$, $\mathbf{A}^\dagger$, and $N_a^\dagger$. Let $\mathbf{A}^H = \mathbf{A}^\dagger$ and $N_a = N_a^\dagger$.
8:
9: **if** $N_a^H = |\mathcal{D}^H|$ **then**
10:     Update $\hat{m}_l = \hat{m}_l + 1$, $\forall l$; Update $\mathcal{R}^\dagger = \{l | l \in [1, r^R], \hat{m}_l \leq n_m\}$; For each slot $l \in \mathcal{R}^\dagger$ and any $l' \in \{r^H + l, 2r^H + l, \ldots, r^R - r^H + l\}$, add l' to $\mathcal{R}^\dagger$ and let $\Gamma_{\hat{m},l'}$ equal $\Gamma_{\hat{m},l}$. Then, calculate $\tau_{\hat{m},l}$, $\forall l \in \mathcal{R}^\dagger$.
11:
12:     Run Algorithm 1 with inputs $\{\Gamma_{\hat{m},l}\}_{\forall l}$ and $\mathcal{R}^\dagger$ from Step 10, $\mathcal{D}^\dagger = \mathcal{D}^R$, $T_f^\dagger = T_f^R$, $r^\dagger = r^R$, $\tau_0^\dagger = \tau_0^R$, $\delta^\dagger = \delta^R$, $\rho^\dagger = \rho^R$. Obtain output $\{\hat{m}_l\}_{\forall l}$, $\{\Gamma_{\hat{m},l}\}_{\forall l}$, $\mathbf{A}^\dagger$, and $N_a^\dagger$. Let $\mathbf{A}^R = \mathbf{A}^\dagger$ and $N_a = N_a + N_a^\dagger$.
13:
14:     **if** $N_a^\dagger = |\mathcal{D}^R|$ **then**
15:         Update $\hat{m}_l = \hat{m}_l + 1$, $\forall l$; Update $\mathcal{R}^\dagger = \{l | l \in [1, r^L], \hat{m}_l \leq n_s\}$; For each slot $l \in \mathcal{R}^\dagger$ and any $l' \in \{r^R + l, 2r^R + l, \ldots, r^L - r^R + l\}$, add l' to $\mathcal{R}^\dagger$ and let $\Gamma_{\hat{m},l'}$ equal $\Gamma_{\hat{m},l}$. Then, calculate $\tau_{\hat{m},l}$, $\forall l \in \mathcal{R}^\dagger$.
16:         Run Algorithm 1 with inputs $\{\Gamma_{\hat{m},l}\}_{\forall l}$ and $\mathcal{R}^\dagger$ from Step 15, $\mathcal{D}^\dagger = \mathcal{D}^L$, $T_f^\dagger = T_f^L$, $r^\dagger = r^L$, $\tau_0^\dagger = \tau_0^L$, $\delta^\dagger = \delta^L$, $\rho^\dagger = \rho^L$. Obtain output $\mathbf{A}^\dagger$, and $N_a^\dagger$. Let $\mathbf{A}^L = \mathbf{A}^\dagger$ and $N_a = N_a + N_a^\dagger$.
17:
18:         Set $F_s = 1$ if $N_a = D$.
19:     **end if**
20: **end if**
21: **Return** $\mathbf{A} = [\mathbf{A}^H, \mathbf{A}^R, \mathbf{A}^L]$, F_s.

- Step 16 of Algorithm 1 and Steps 10 and 15 of Algorithm 2—These steps move from the current mini-slot to the next mini-slot of the same slot. As a result, the AD-F of the next mini-slot needs to be calculated. The details regarding the calculation of $\tau_{\hat{m},l}$ in these steps can be found in [57];
- Step 2 of Algorithm 2—Since each LP assignment cycle consists of r^L / r^H HP cycles and r^L / r^R RP cycles, respectively, the HP and RP assignment cycles can be found accordingly after obtaining the LP cycle length based on (2.15);

- Step 21 of Algorithm 2—The element in the first/second row and the ith column of the device assignment matrix, $\mathbf{A}$, gives the index of the slot/mini-slot assigned to device i;
- Matrix $\mathbf{A}$ only gives the first slot/mini-slot assigned to device i. If device i is an HP device and assigned slot and mini-slot $\{l, m\}$, then it is also assigned slot/mini-slot $\{l', m\}$ for any $l' \in \{r^H + l, 2r^H + l, \ldots, r^L - r^H + l\}$. If device i is an RP device and assigned slot and mini-slot $\{l, m\}$, then it is also assigned slot/mini-slot $\{l', m\}$ for any $l' \in \{r^R + l, 2r^R + l, \ldots, r^L - r^R + l\}$. This is reflected in Steps 10 and 15 of Algorithm 2 and consistent with the illustration in Fig. 2.6.

In the core assignment algorithm (Algorithm 1), adding a device to a mini-slot has an impact on $\Lambda_{m,l}$, $\Gamma_{m,l}$, and $q^c_{m,l}$. Therefore, after assigning device i mini-slot m of slot l, these variables need to be updated for the mini-slot. If device i is the first device assigned mini-slot m of slot l, the following update applies:

$$\tilde{q}^c_{m,l} = 0 \tag{2.16a}$$

$$\tilde{n}_i = 1, \tag{2.16b}$$

$$\tilde{\Lambda}_{m,l} = \lambda_i \tag{2.16c}$$

$$\tilde{\Gamma}_{m,l} = \Gamma_{m,l} + T_f^\dagger \lambda_i \tag{2.16d}$$

$$\tilde{\tau}_{m,l} = \tau_{m,l} \tag{2.16e}$$

where $\tilde{x}$ represents an updated value of x after assigning device i, and $T_f^\dagger$ is the corresponding (HP, RP, or LP) cycle length. If device i is not the first device assigned mini-slot m of slot l, the following update applies:

$$\tilde{q}^c_{m,l} = \left(1 - (1 - q^c_{m,l})(1 - T_f^\dagger \lambda_i)\right) \tag{2.17a}$$

$$n^c_i = 1 + \sum_{j \in \mathcal{D}_{m,l}\setminus\{i\}} \tau_{m,l} T_f^\dagger \lambda_j \tag{2.17b}$$

$$\tilde{\Lambda}_{m,l} = \Lambda_{m,l} + \lambda_i \left(1 - \frac{\tilde{q}^c_{m,l}}{n^c_i}\right) \tag{2.17c}$$

$$\tilde{\Gamma}_{m,l} = \Gamma_{m,l} + T_f^\dagger \lambda_i \left(1 - \frac{\tilde{q}^c_{m,l}}{n^c_i}\right) \tag{2.17d}$$

$$\tilde{\tau}_{m,l} = \tau_{m,l} \tag{2.17e}$$

which is based on the analysis in Sect. 2.4.5 of Sect. 2.4. Equations (2.17a)–(2.17d) update the packet collision probability,[14] the average number of packets per transmission (taking collision into account), the aggregated packet arrival rate, and the accumulated number of packet arrivals, respectively, corresponding to a mini-slot after a new device is assigned that mini-slot. The last equation, i.e., (2.17e), follows from [57] (the proof of Theorem 3 therein). Specifically, the result shows that, under a low collision probability, the AD-F for devices assigned any mini-slot depends on the packet arrival rates of all devices in the preceding mini-slots, but not the packet arrival rates of other devices sharing the same mini-slot.

2.5.4 AI-Assisted Protocol Parameter Selection

The proposed device assignment in the preceding section can be applied when parameters n_m, r^H, r^R, and r^L are given. In this subsection, we propose learning-assisted scheduling to determine the values of these protocol parameters.

The motivation for learning-assisted scheduling roots in the complexity of choosing proper values for the protocol parameters. First, the impact of protocol parameters n_m, r^H, r^R, r^L and the impact of device assignment are correlated. For example, knowledge of the slot/mini-slot assignment is required to analyze the impact of n_m, while the assignment cannot be determined without knowing n_m first. Second, the effects of n_m, r^H, r^R, r^L on the performance are mutually dependent. Consider n_m and r^H as an example. Both n_m and r^H affect the delay of HP devices. The impact of adjusting r^H depends on the value of n_m, and the dependence is further affected by the device packet arrival rate profile. As a result, we cannot establish an analytical model for determining n_m, r^H, r^R, and r^L. On the other hand, using brutal force to choose their values is not viable due to the large number of diverse devices. There are usually too many candidate combinations of n_m, r^H, r^R, and r^L, and each combination requires a re-calculation of the device assignment using Algorithms 1 and 2. As the assignment algorithm is based on calculating the delay and collision probability while assigning each device, the complexity of recalculating all assignment for all combinations can be very high.[15]

To cope with the complexity of scheduling problem, we use a learning-based method to capture the impact of n_m, r^H, r^R, r^L and determine their values. Specifically, we train a DNN to learn the mapping from the combination of device and packet arrival rate profiles and protocol parameter settings to the protocol performance. A significant part of the training can be done offline to avoid a long

---

[14] In practice, a guard margin may need to be applied to the estimated collision probability in (2.17a). After all, such estimation may not be sufficiently accurate since we assume no statistical knowledge of the packet arrival of any device other than the average arrival rate.

[15] Such complexity, as the result of a mixed integer nonlinear programming, is noted in many works, e.g., [79], some of which adopt a learning-based method.

training duration in an online setting caused by searching for and determining appropriate protocol parameters. The DNN assists in determining the parameters of the proposed MAC protocol as follows. First, for each device and packet arrival rate profile,[16] we try different combinations of n_m, r_H, r_R, and r_L, use the heuristic algorithm to obtain the assignment, and test the resulting performance using simulations. Then, the device and packet arrival rate profile, protocol parameter setting (n_m, r_H, r_R, and r_L), and the resulting protocol performance (as label) are used to train and test the DNN.

The data generation, training, and testing are conducted offline. When the DNN is well-trained, we can imitate the mapping from a device and packet arrival rate profile and a protocol parameter setting to the protocol performance. Accordingly, we can determine the protocol parameters online by trying different parameters on the DNN and compare the resulting performance. Recall that the packet arrival rates of devices remain constant in a relatively long duration, as mentioned in Sect. 2.3. When an update of the packet arrival rates occurs, it triggers a decision on the protocol parameters, and the DNN assists the decision making as aforementioned.

The input of the DNN includes the following two components:

- Device and packet arrival rate profile—To be flexible with the number of devices, we divide the range of packet arrival rate into I intervals. Letting λ^{max} and λ^{min} denote the maximum and minimum packet arrival rates, the width of each interval is $(\lambda^{max} - \lambda^{min})/I$. We count the number of HP, RP, and LP devices in each of the I intervals and organize the corresponding numbers into three $I \times 1$ vectors c^H, c^R, and c^L, respectively;
- Protocol parameter settings—The number of mini-slots in each slot (n_m) and the number of slots in each HP, RP, and LP assignment cycle (r^H, r^R, and r^L) are the second input component.

The input data, $\{c^H, c^R, c^L, n_m, r^H, r^R, r^L\}$, is normalized by the Z-score method [80] and fed to the first fully connected layer.

The DNN consist of K fully connected layers. For layer k, n_k neurons are deployed. The trainable parameters, i.e., kernels and bias, for neurons in the network are denoted by θ. The DNN output includes the maximum and the average delay as well as the maximum and the average packet collision probability for each of the three device types. In addition, we adopt an indication bit in the output to indicate whether the assignment algorithms fail to find a solution that satisfies the performance requirements of all devices. The indication bit is 1 if the assignment attempt fails and 0 otherwise. Overall, there are 13 output neurons introduced in the network.

The DNN following the above-mentioned design is illustrated in Fig. 2.7. The DNN is implemented by Keras [81], a high-level neural network application programming interface using Tensorflow backend. The objective of the offline

---

[16] We refer to the collective information including the number of HP, RP, and LP devices as well as the packet arrival rate of each device as a device and packet arrival rate profile.

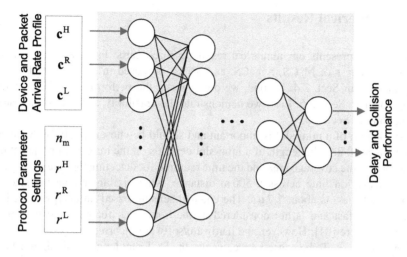

Fig. 2.7 The structure of the DNN

training is to find an appropriate θ value that minimizes the loss function $\mathcal{L}(\theta)$ represented by the mean squared error (MSE) for regression.

Adam optimizer [82] is adopted to minimize the loss function iteratively, where the optimizer is set with learning rate $\alpha = 1e-3$ and exponential decay rates $\beta_1 = 0.9$ and $\beta_2 = 0.999$.

The labels, i.e., the protocol performance under specific device and packet arrival rate profiles and the protocol parameter settings, are generated via simulations. Although we can generate the labels offline, a very large training set may not be practical as it could require overwhelmingly long simulations. Meanwhile, the simulation results also demonstrate randomness, due to the randomness in the packet arrival at each device. Given the limited training set with randomness in the labels, the problem of over-fitting can be severe. We can use random dropout to alleviate over-fitting and improve the robustness of the training model [83].

It is worth noting that our DNN does not directly output the best protocol parameters $\{n_m, r^H, r^R, r^L\}$. An alternative design is to train a DNN that outputs the best $\{n_m, r^H, r^R, r^L\}$. The difference is whether the DNN assists the decision making or directly makes a decision. We choose the former and let the DNN learns the mapping from various protocol parameters to the resulting performance since this approach is more flexible. For example, if the DNN directly makes a decision, the output may not be feasible or preferred when there are additional constraints on $\{n_m, r^H, r^R, r^L\}$. By contrast, using our approach, we can identify different parameter sets and compare them for a feasible or preferred solution.

2.6 Numerical Results

This section presents our numerical results in three parts. First, we demonstrate
the effectiveness of MsCS, SyncCS, and SMsA proposed in Sect. 2.3 and verify
our analysis in Sect. 2.4. Second, we demonstrate the performance of the device
assignment in Sect. 2.5. Last, we demonstrate the feasibility of the DNN-assisted
scheduling in Sect. 2.5.

The length of a mini-slot is important and should be chosen carefully. As mentioned in Sect. 2.3, the length of a mini-slot depends on the maximum propagation
delay across the coverage area and the time required for detecting the channel status.
The propagation time across a 500 m distance, which is larger than the size of
typical factories, is about 1.7 μs. The channel sensing based on energy detection
can be very fast and is not considered as the bottleneck for reducing the mini-
slot length here [84]. However, the hardware/software incurred delay can vary for
different devices. To be conservative, we use the DCF slot length in IEEE 802.11ac
as the reference and set the mini-slot length to be 9 μs in most of our simulation
examples [85]. Using this mini-slot length, the overhead in each slot incurred by
having n_m mini-slot for channel sensing is $9 \times 10^{-6} \times n_m$ seconds. For example,
consider a packet length of 50 bytes in the physical layer, and a data transmission
rate of 3 Mb/s, which yields a data transmission duration of 133 μs. With 10 mini-
slots in each slot, the overall length of mini-slots is 90 μs in every 223 μs.

2.6.1 Mini-Slot Delay with MsCS, SyncCS, and SMsA

Via simulations, we evaluate the mini-slot delay[17] in the cases with and without
SyncCS and SMsA and compare the numerical results with the analytical results
from Sect. 2.4. We focus on different mini-slots of one target slot. The general
settings in this subsection are as follows (unless stated otherwise):

- n_m and n_s are set to 10 and 100, respectively;
- T_m is set to 9 μs. T_x is 133 μs, i.e., the duration of a 50-byte physical-layer packet
  transmitting at 3 Mb/s. Accordingly, T_s in its full length is 223 μs, i.e., 10×9 μs
  + 133 μs;
- Device i is assigned a mini-slot with smaller index than the mini-slot of device j
  if $\lambda_i < \lambda_j$;
- 20,000 frames are simulated for each case.

Mini-slot delay with MsCS and with both MsCS and SyncCS: Fig. 2.8 shows the
results with only MsCS (i.e., no SyncCS or SMsA), with and without buffer, as well
as the results with both MsCS and SyncCS, in the case with buffer, for Poisson

---

[17] For brevity, we use 'mini-slot delay' to refer to the delay of a device assigned that mini-slot.

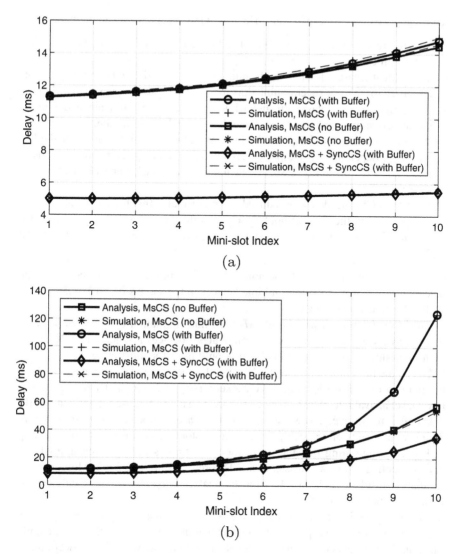

Fig. 2.8 Mini-slot delay of MsCS only and of MsCS and SyncCS with (**a**) 0.2 to 1 packets per second per device; (**b**) 1 to 5 packets per second per device

packet arrivals. The overall delay includes both the base and the access delay. The packet arrival rate of each device is randomly generated based on a uniform distribution. Figure 2.8a corresponds to a lower packet arrival rate, i.e., in the range between 0.2 and 1 packets per second per device, and Fig. 2.8b corresponds to a higher packet arrival rate, i.e., between 1 and 5 packets per second per device. The analytical results in Fig. 2.8 are based on (2.3) and (2.6) with the expected frame length given by (2.11), respectively. The following observations can be made:

- The difference between the analytical and numerical results is small for all mini-slots in all cases;
- The delay increases slowly with the mini-slot index for the first several mini-slots but faster for the last several mini-slots in the case of higher packet arrival rate;
- The difference in delay with and without buffer is insignificant under lower packet arrival rate and significant under higher packet arrival rate;
- Without SyncCS, the delay for the first mini-slot is around 11 ms. For the last mini-slot, depending on the packet arrival rate, the delay ranges from 15 ms in Fig. 2.8a to 125 ms in Fig. 2.8b, less than the average packet arrival interval in all cases;
- With SyncCS, the delay is reduced by more than 50% for each mini-slot as compared with the case without SyncCS. In the case of a higher packet arrival rate in Fig. 2.8b, the maximum delay decreases from about 125 ms to around 35 ms.

Overall, the numerical results demonstrate the accuracy of (2.3) and (2.6), the practicality of accommodating multiple devices in the same slot via MsCS, as well as the effectiveness of SyncCS.

Mini-slot delay with MsCS and SMsA: In this simulation example with SMsA (but not SyncCS), each mini-slot accommodates 7 devices instead of one. Note that such mini-slot usage is not optimal and is only used for illustrating the impact of SMsA on the mini-slot delay. As the 10 mini-slots accommodate 70 devices in total, the slot is prone to overloading if the packet arrival rate is high. Therefore, we use low packet arrival rate in this example. Figure 2.9 shows the case (a) without and (b) with buffer, respectively. Now that each mini-slot accommodates 7 devices, there are 7 numerical results on the delay for each mini-slot. The simulation results overlap in Fig. 2.9, suggesting that the delay for all 7 devices in any given mini-slot is almost identical. This is consistent with Theorem 2.3 in Sect. 2.4.5. Moreover, the simulation results match closely with the analytical results (some details can be found in Appendix C of [57]).

Impact of mini-slot length and frame length: We use the same settings as in Fig. 2.9 with buffers, except for a change in the mini-slot length or the frame length. The mini-slot usage here is still not optimal and only for showing the impact of mini-slot and frame lengths. In Fig. 2.10a, the mini-slot length reduces to 7 µs from 9 µs in Figs. 2.8 and 2.9. Comparing with Fig. 2.9b, the impact of mini-slot length on the delay becomes evident. Accordingly, the performance of the proposed protocol can further improve if a reduction in the mini-slot length is feasible. In Fig. 2.10b, the mini-slot length is back to 9 µs, the packet arrival rate is multiplied by 5, and the frame length reduces to 5 slots from 100 slots. Comparing with Fig. 2.9b, the impact of frame length on the delay and the necessity of differentiated assignment cycles become clear. The results indicate that a very low delay is achievable if we keep the HP assignment cycle sufficiently short.

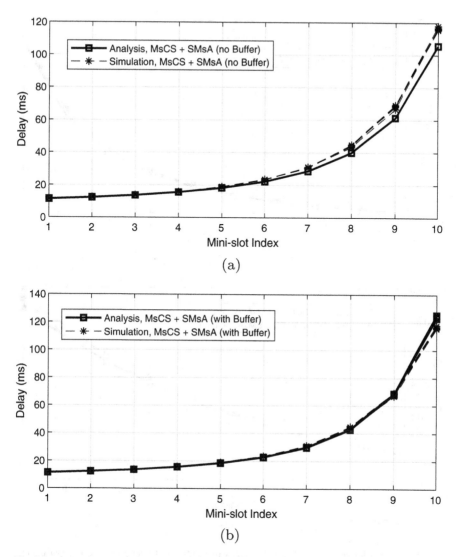

Fig. 2.9 Mini-slot delay of MsCS and SMsA with (**a**) 0.2 to 1 packets per second per device, no buffer; (**b**) 0.2 to 1 packets per second per device, with buffer. There are 7 overlapping dashed curves in each plot, corresponding to the simulation results. Given any mini-slot index, the 7 points on the 7 dashed curves are for the 7 devices sharing the corresponding mini-slot. The only solid curve in each plot gives the analytical result for all devices, since Theorem 2.3 suggests that the delay for all devices sharing the same mini-slot is approximately the same

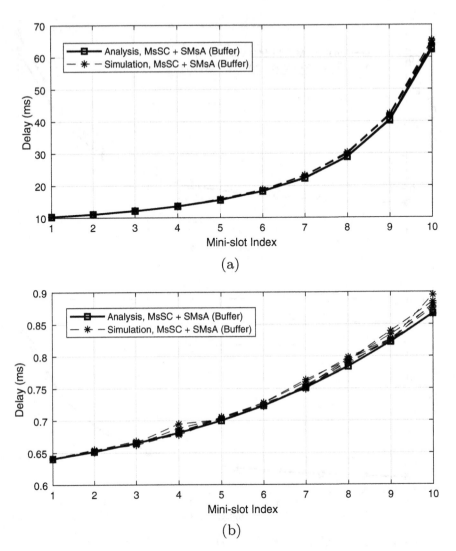

Fig. 2.10 Mini-slot delay of MsCS and SMsA with (**a**) 0.2 to 1 packets per second per device, $7\mu s$ mini-slot, 100 slots per frame; (**b**) 1 to 5 packets per second per device, $9\,\mu s$ mini-slot, 5 slots per frame. The 7 overlapping dashed curves in each plot are the result of 7 devices sharing each mini-slot. The only solid curve in each plot gives the analytical result for all devices based on Theorem 2.3

2.6.2 Performance of the Device Assignment Algorithms

We evaluate the performance of the device assignment, i.e., Algorithms 1 and 2 in Sect. 2.5, given n_m, r^H, r^R, and r^L. In the evaluation, MsCS, SyncCS, SMsA, as well as differentiated assignment cycles are used, and a buffer is assumed at each device. Again, T_m and T_x are set as $9\,\mu s$ and $133\,\mu s$, respectively.

We consider 1000 devices with mixed packet arrival patterns. Specifically, the number of HP, RP, and LP devices is 50, 450, and 500, respectively. A half of all the devices, selected randomly, have Poisson packet arrivals with rate randomly selected from the range between 1 packet per second per device and 5 packets per second per device. The remaining devices have periodic packet arrivals. The arrival rate is randomly distributed in the same range (i.e., $[1, 5]$), and a random component within $\pm 5\%$ of the packet arrival interval is added to each arrival instant for periodical packets. Each slot consists of 8 mini-slots (i.e., $n_m = 8$), and each HP assignment cycle consists of 5 slots (i.e., $r^H = 5$). Delay thresholds δ^H, δ^R, δ^L are set to 1ms, 10ms, and 80ms, respectively, while the packet collision probability thresholds ρ^H, ρ^R, and ρ^L are set to 1.5%, 6%, and 10%, respectively.

A simulation duration of 2000 seconds is used to test the performance of Algorithms 1 and 2. Figure 2.11 shows the delay and packet collision probability of each device as well as the average for each type of devices, with two different assignment cycle settings. The three clusters in each figure correspond to the three groups of HP, RP, and LP devices, respectively. In Fig. 2.11a, r^R and r^L are 45 and 270, respectively, while r^R and r^L are 35 and 140 in Fig. 2.11b. From Fig. 2.11, we observe that the preset QoS requirements for all devices are satisfied. For example, from Fig. 2.11a, the following observations can be made:

- HP devices—average delay 0.38 ms, maximum delay 0.39 ms; average collision probability 0.54%, and maximum collision probability 1.08%;
- RP devices—average delay 3.1 ms, maximum delay 3.7 ms, average collision probability 1.4%, and maximum collision probability 4.8%;
- LP devices—average delay 14.2 ms, maximum delay 20.9 ms, average collision probability 0%, and maximum collision probability 0%.

Figure 2.11 also clearly demonstrates differentiated performance achieved for different type of devices. Note that the delay in Fig. 2.11 is smaller than that in Figs. 2.8 and 2.9 for two reasons. First, differentiated assignment cycles enable a very low delay for HP and RP devices. For example, each HP device gets a potential transmission opportunity in every 5 slots in the case of Fig. 2.11, the same as in Fig. 2.10b, instead of every 100 slots in the case of Figs. 2.8 and 2.9. Second, each slot consists of only 8 mini-slots in the case of Fig. 2.11, instead of 10 in the case of Figs. 2.8 and 2.9. A less number of mini-slots leads to both shorter slots, which reduce delay for all devices, and higher slot idle probabilities, which contribute to a further reduction in delay thanks to SyncCS.

Further, Fig. 2.11 shows the impact of assignment cycles on the performance. Specifically, via different settings of r^R and r^L in Fig. 2.11a and b, the possibility

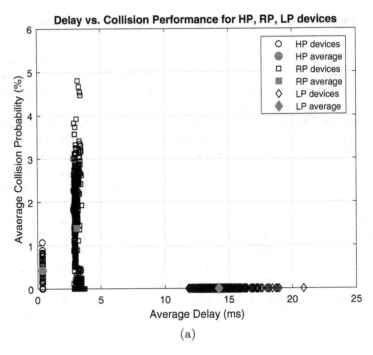

(a)

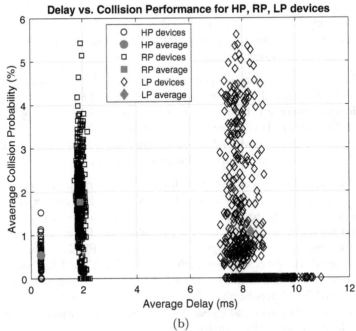

(b)

Fig. 2.11 The performance of Algorithms 1 and 2 with 1000 devices and mixed packet arrival patterns: (**a**) Delay and collision performance, $r^R = 45$, $r^L = 270$; (**b**) delay and collision performance, $r^R = 35$, $r^L = 140$

of making a trade-off between collision and delay is shown. Moreover, Fig. 2.11a and b demonstrate how our proposed algorithms can adapt to the given protocol parameters. In Fig. 2.11a, r^L is larger and each LP device has to wait for a longer duration before having a transmission opportunity. As a result, the probability that an LP device has a packet to send in its assigned mini-slot can be high, and assigning two or more LP devices the same mini-slot in such case can yield a high collision probability. Therefore, the algorithms choose to assign each LP device an exclusive mini-slot. In comparison, r^L is much smaller in Fig. 2.11b, and thus the probability that an LP device has a packet to send in its assigned mini-slot is lower. Therefore, the algorithms allow LP devices to share a mini-slot at the cost of small collision probabilities.

Figure 2.12 demonstrates the performance under the same setting as in Fig. 2.11 except: (1) there are now 350 devices, all HP, in the network; and (2) there are 4 mini-slots in each slot ($n_m = 4$) and 6 slots in each HP cycle ($r^H = 6$). The QoS requirements on delay and packet collision are satisfied for all devices. The average delay and collision probability among all devices are less than 0.26ms and 0.6%, respectively. This result illustrates the flexibility of the proposed device assignment algorithms in terms of adapting to different device profiles.

In the simulation examples in this subsection, the number of mini-slots, n_m, and the assignment cycles, r^H, r^R, and r^L, are not optimized. Thus, the resulting performance is not necessarily optimal. However, the results shown in Fig. 2.11 illustrate the advantage of the proposed MAC protocol and the assignment algorithms, in

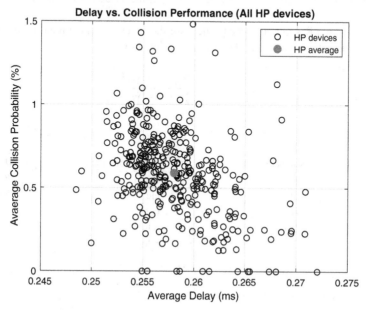

Fig. 2.12 The performance of Algorithms 1 and 2 with 350 HP devices, $n_m = 4$, $r^H = 6$

terms of satisfying stringent QoS, prioritization, and flexibility. Particularly, while random access is known to have distinctive advantage for low data traffic in delay as compared with scheduled access, e.g., as discussed in [86], we demonstrate that appropriate scheduling, combined with well-designed access protocol, can also achieve very low delay in a high-density MTC network.

2.6.3   DNN-Assisted Scheduling

The structure parameters of our proposed DNN are given in Table 2.2. We utilize 8200 sets of device packet arrival profiles and generate the corresponding delay and packet collision performance via the device assignment algorithms, for various values of n_m and r^R.[18] Each of the 8200 sets consists of 6 different combinations of n_m and r^R, yielding 49,200 data entries. We employ 80% of 49,200 data entries as the training set, 10% as the validation set in training, and 10% as the test set. To deal with the overfitting issue in training, we utilize the random dropout technique. Specifically, the neurons in layers n_1 and n_2 have a 70% chance to be dropped off in each training step. The gradient backpropagation is performed over data batches of size 128 during 50 epochs.

The training loss and validation loss of the proposed DNN are shown in Fig. 2.13a, where the output data are normalized to the range [0, 1]. The convergence occurs after around 20 epochs. In addition, the gap between training loss and validation loss is small, showing that the overfitting issue is alleviated by random dropout.

We adopt the R-squared score to measure the fitness of our trained model in the training data set. The R-squared score is calculated by

$$\text{R-square} = \frac{\sum_i (\hat{y}_i - \bar{y}_i)^2}{\sum_i (y_i - \bar{y}_i)^2}. \tag{2.18}$$

Table 2.2 DNN structure

Layer	Number of neurons	Activation function	Dropout
n_1	1024	elu	70%
n_2	1024	elu	70%
n_3	512	elu	–
n_4	256	relu	–
n_5	128	relu	–
n_6	64	relu	–
n_7	13	relu	–

---

[18] We fix r^H and r^L in this illustration for simplicity.

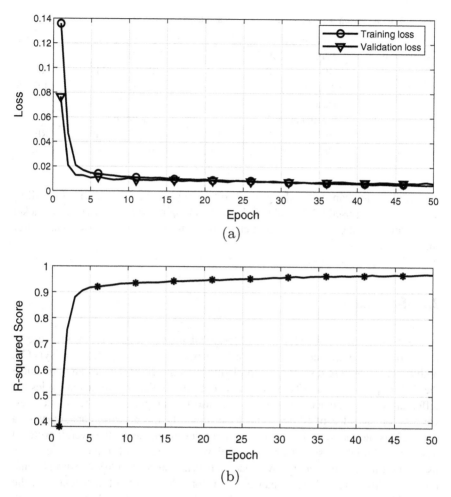

Fig. 2.13 (a) Training loss and validation loss of the proposed DNN; (b) R-squared score of the proposed DNN

When the score is close to 1, the trained model can generate predicted results with a reasonably small variance. The R-squared score of the proposed DNN is shown in Fig. 2.13b, in which the score converges to a value close to 1 after 10 epochs.

We further validate the fitness of the trained DNN model with the data from the test set. The comparison between the predicted performance metric values and the ground truth labels is presented in Table 2.3.[19] It can be seen that the

---

[19] The LP devices always have 0 collision probability in this example (similar to the case in Fig. 2.11a. Thus, the MSE is 0 but not meaningful in such cases. Therefore, we use two '-' under LP instead of '0' in this table.

Table 2.3 Comparison between predicted results and labels in the test set

Overall MSE	Collision probability					
	Maximum MSE			Mean MSE		
	HP	LP	RP	HP	LP	RP
2.3e−4	2.9e−5	–	1.8e−4	1.2e−6	–	1.6e−5
Flag bit accuracy	Delay					
	Maximum MSE			Mean MSE		
	HP	LP	RP	HP	LP	RP
98.5%	5.8e−9	4.0e−5	2.6e−7	5.0e−9	1.4e−5	1.5e−7

predicted results can match the ground truth labels in the test set with low MSE, and thus the proposed DNN is able to learn the mapping from the device and packet arrival profile and the protocol parameter settings to the resulting performance after sufficient training.

2.7   Summary

In this chapter, we first tailor a MAC protocol for the smart factory use case in IIoT. To increase channel utilization efficiency, we propose MsCS and SyncCS, both of which feature distributed coordination. To prioritize devices and guarantee the QoS requirement of HP devices, we adopt differentiated assignment cycles for different types of devices. To further increase the supported number of devices, we develop the idea of SMsA, which can multiply the network capacity with a delay-collision trade-off. Thanks to the above design elements, the overall protocol has the potential to simultaneously achieve the targets of improving channel usage, minimizing messaging overhead, satisfying stringent QoS constraints, and providing differentiated performance. Then, for achieving the full potential of the proposed protocol, we customize scheduling for our proposed MAC protocol to complete the overall connectivity solution. Based on the performance analysis, we are able to assign devices with the due granularity and accuracy. Utilizing a trained DNN, we manage to determine the protocol parameters efficiently. Integrating the distributed coordination and the centralized scheduling composes the unique strength of our tailored MAC design. As a result, the proposed MAC is capable of supporting a large number of devices with sporadic data packets under a single AP and a single channel, while achieving a (sub)millisecond-level delay and very low collision probability.

References

1. Aceto, G., Persico, V., Pescapé, A.: A survey on information and communication technologies for industry 4.0: State-of-the-art, taxonomies, perspectives, and challenges. IEEE Commun. Surv. Tut. **21**(4), 3467–3501 (2019)
2. Beecks, C., Grass, A., Devasya, S., Jentsch, M., Soto, J.A.C., Tavakolizadeh, F., Linnemann, A., Eisenhauer, M.: Smart data and the industrial Internet of Things. In: Next Generation Internet of Things: Distributed Intelligence at the Edge and Human Machine-to-Machine Cooperation, pp. 175–206. River Publishers, Denmark (2018)
3. Industrial IoT (IIoT) market by component, application (robotics, maintenance, monitoring, resource optimization, supply chain, management), industry (aerospace, automotive, energy, healthcare, manufacturing, retail), and geography - global forecast to 2027. Tech. Rep. ID: MRICT-104404, Meticulous Market Research (2020)
4. Vitturi, S., Zunino, C., Sauter, T.: Industrial communication systems and their future challenges: next-generation Ethernet, IIoT, and 5G. In: Proceedings of the IEEE
5. Service requirements for machine-type communications (MTC). Tech. Rep. TS 22.368, Version 16.0.0, Release 16, 3GPP (2020)
6. Bockelmann, C., Pratas, N., Nikopour, H., Au, K., Svensson, T., Stefanovic, C., Popovski, P., Dekorsy, A.: Massive machine-type communications in 5G: physical and MAC-layer solutions. IEEE Commun. Mag. **54**(9), 59–65 (2016)
7. Sharma, S.K., Wang, X.: Toward massive machine type communications in ultra-dense cellular IoT networks: current issues and machine learning-assisted solutions. IEEE Commun. Surv. Tut. **22**(1), 426–471 (2020)
8. Sisinni, E., Saifullah, A., Han, S., Jennehag, U., Gidlund, M.: Industrial Internet of Things: challenges, opportunities, and directions. IEEE Trans. Ind. Inf. **14**(11), 4724–4734 (2018)
9. Brown, G.: Ultra-reliable low-latency 5G for industrial automation. Tech. rep. (2018)
10. 5G for connected industries and automation. Tech. Rep. 2nd Edition, 5G Alliance for Connected Industries and Automation (2019)
11. Study on self evaluation towards IMT-2020 submission. Tech. Rep. TR 37.910, Version 1.0.0, 3GPP (2018)
12. Study on communication for automation in vertical domains. Tech. Rep. TS 22.804, Version 16.3.0, Release 16, 3GPP (2020)
13. Popli, S., Jha, R.K., Jain, S.: A survey on energy efficient narrowband Internet of Things (NBIoT): architecture, application and challenges. IEEE Access **7**, 16739–16776 (2019)
14. Xia, N., Chen, H.H., Yang, C.S.: Emerging technologies for machine-type communication networks. IEEE Netw. **34**(1), 214–222 (2020)
15. LTE-M performance towards 5G IoT requirements. Tech. Rep. V1.1, Sierra Wireless et al. (2019)
16. Dawaliby, S., Bradai, A., Pousset, Y., Chatellier, C.: Joint energy and QoS-aware memetic-based scheduling for M2M communications in LTE-M. IEEE Trans. Emerg. Top. Comput. Intell. **3**(3), 217–229 (2019)
17. Technical specification group services and system aspects; release 15 description; summary of rel-15 work items (release 15). Tech. Rep. TR 21.915, version 15.0.0, 3GPP (2019)
18. Saad, W., Bennis, M., Chen, M.: A vision of 6G wireless systems: applications, trends, technologies, and open research problems. IEEE Netw. **34**(3), 134–142 (2020)
19. Adelantado, F., Vilajosana, X., Tuset-Peiro, P., Martinez, B., Melia-Segui, J., Watteyne, T.: Understanding the limits of LoRaWAN. IEEE Commun. Mag. **55**(9), 34–40 (2017)
20. El Rachkidy, N., Guitton, A., Kaneko, M.: Collision resolution protocol for delay and energy efficient LoRa networks. IEEE Trans. Green Commun. Netw. **3**(2), 535–551 (2019)
21. Ali, M.Z., Mišić, J., Mišić, V.B.: Performance evaluation of heterogeneous IoT nodes with differentiated QoS in IEEE 802.11ah RAW mechanism. IEEE Trans. Veh. Technol. **68**(4), 3905–3918 (2019)

22. Adame, T., Bel, A., Bellalta, B., Barcelo, J., Oliver, M.: IEEE 802.11ah: the WiFi approach for M2M communications. IEEE Wireless Commun. **21**(6), 144–152 (2014)
23. Mahesh, M., Pavan, B.S., Harigovindan, V.: Data rate-based grouping to resolve performance anomaly of multi-rate IEEE 802.11ah IoT networks. IEEE Netw. Lett. **2**(4), 166–170 (2020)
24. Chang, T.C., Lin, C.H., Lin, K.C.J., Chen, W.T.: Traffic-aware sensor grouping for IEEE 802.11ah networks: regression based analysis and design. IEEE Trans. Mobile Comput. **18**(3), 674–687 (2019)
25. Yang, G., Xiao, M., Poor, H.V.: Low-latency millimeter-wave communications: Traffic dispersion or network densification? IEEE Trans. Commun. **66**(8), 3526–3539 (2018)
26. Chen, S., Zhao, T., Chena, H.H., Meng, W.: Network densification and path-loss models vs. UDN performance–a unified approach. IEEE Trans. Wireless Commun. **20**, 4058–4071 (2021)
27. Shen, X., Gao, J., Wu, W., Lyu, K., Li, M., Zhuang, W., Li, X., Rao, J.: AI-assisted network-slicing based next-generation wireless networks. IEEE Open J. Veh. Technol. **1**, 45–66 (2020)
28. Ye, Q., Li, J., Qu, K., Zhuang, W., Shen, X.S., Li, X.: End-to-end quality of service in 5G networks: examining the effectiveness of a network slicing framework. IEEE Veh. Technol. Mag. **13**(2), 65–74 (2018)
29. Ma, G., Ai, B., Wang, F., Zhong, Z.: Joint design of coded tandem spreading multiple access and coded slotted ALOHA for massive machine-type communications. IEEE Trans. Ind. Inf. **14**(9), 4064–4071 (2018)
30. Xie, R., Yin, H., Chen, X., Wang, Z.: Many access for small packets based on precoding and sparsity-aware recovery. IEEE Trans. Commun. **64**(11), 4680–4694 (2016)
31. Wang, B., Dai, L., Yuan, Y., Wang, Z.: Compressive sensing based multi-user detection for uplink grant-free non-orthogonal multiple access. In: Proceedings of the 2015 IEEE 82nd Vehicular Technology Conference (VTC2015-Fall), pp. 1–5 (2015)
32. Moussa, H.G., Zhuang, W.: RACH performance analysis for large-scale cellular IoT applications. IEEE Internet Things J. **6**(2), 3364–3372 (2019)
33. Yuan, J., Huang, A., Shan, H., Quek, T.Q., Yu, G.: Design and analysis of random access for standalone LTE-U systems. IEEE Trans. Veh. Technol. **67**(10), 9347–9361 (2018)
34. Abbas, R., Shirvanimoghaddam, M., Li, Y., Vucetic, B.: Random access for M2M communications with QoS guarantees. IEEE Trans. Commun. **65**(7), 2889–2903 (2017)
35. Bui, A.T.H., Nguyen, C.T., Thang, T.C., Pham, A.T.: A comprehensive distributed queue-based random access framework for mMTC in LTE/LTE-A networks with mixed-type traffic. IEEE Trans. Veh. Technol. **68**(12), 12,107–12,120 (2019)
36. Leyva-Mayorga, I., Tello-Oquendo, L., Pla, V., Martinez-Bauset, J., Casares-Giner, V.: On the accurate performance evaluation of the LTE-A random access procedure and the access class barring scheme. IEEE Trans. Wireless Commun. **16**(12), 7785–7799 (2017)
37. Wali, P.K., Das, D.: Optimization of barring factor enabled extended access barring for energy efficiency in LTE-advanced base station. IEEE Trans. Green Commun. Netw. **2**(3), 830–843 (2018)
38. Vidal, J.R., Tello-Oquendo, L., Pla, V., Guijarro, L.: Performance study and enhancement of access barring for massive machine-type communications. IEEE Access **7**, 63,745–63,759 (2019)
39. Lee, M., Kim, Y., Piao, Y., Lee, T.J.: Recycling random access opportunities with secondary access class barring. IEEE Trans. Mobile Comput. **19**(9), 2189–2201 (2020)
40. Jang, H.S., Jin, H., Jung, B.C., Quek, T.Q.S.: Versatile access control for massive IoT: Throughput, latency, and energy efficiency. IEEE Trans. Mobile Comput. **19**(8), 1984–1997 (2020)
41. Pocovi, G., Soret, B., Pedersen, K.I., Mogensen, P.: MAC layer enhancements for ultra-reliable low-latency communications in cellular networks. In: Proceedings of the 2017 IEEE International Conference on Communication Workshops (ICC Workshops), pp. 1005–1010 (2017)
42. You, L., Liao, Q., Pappas, N., Yuan, D.: Resource optimization with flexible numerology and frame structure for heterogeneous services. IEEE Commun. Lett. **22**(12), 2579–2582 (2018)

43. Study on NR industrial Internet of Things (IoT). Tech. Rep. TR 38.825, Version 16.0.0, Release 16, 3GPP (2019)
44. Liu, Y., Deng, Y., Elkashlan, M., Nallanathan, A., Karagiannidis, G.K.: Analyzing grant-free access for URLLC service. IEEE J. Sel. Areas Commun. **39**(3), 741–755 (2021)
45. Park, C.W., Hwang, D., Lee, T.J.: Enhancement of IEEE 802.11ah MAC for M2M communications. IEEE Commun. Lett. **18**(7), 1151–1154 (2014)
46. Nawaz, N., Hafeez, M., Zaidi, S.A.R., McLernon, D.C., Ghogho, M.: Throughput enhancement of restricted access window for uniform grouping scheme in IEEE 802.11 ah. In: Proceedings of the 2017 IEEE International Conference on Communications (ICC), pp. 1–7 (2017)
47. Kim, Y., Hwang, G., Um, J., Yoo, S., Jung, H., Park, S.: Throughput performance optimization of super dense wireless networks with the renewal access protocol. IEEE Trans. Wireless Commun. **15**(5), 3440–3452 (2016)
48. Mahesh, M., Harigovindan, V.: Restricted access window-based novel service differentiation scheme for group-synchronized DCF. IEEE Commun. Lett. **23**(5), 900–903 (2019)
49. Shimokawa, M., Sanada, K., Hatano, H., Mori, K.: Station grouping method for non-uniform station distribution in IEEE 802.11ah based IoT networks. In: Proceedings of the 2020 IEEE 91st Vehicular Technology Conference (VTC2020-Spring), pp. 1–5 (2020)
50. Chang, T.C., Lin, C.H., Lin, K.C.J., Chen, W.T.: Traffic-aware sensor grouping for IEEE 802.11ah networks: regression based analysis and design. IEEE Trans. Mobile Comput. **18**(3), 674–687 (2019)
51. Lei, X., Rhee, S.H.: A novel grouping mechanism for performance enhancement of sub-1 GHz wireless networks. In: Proceedings of the 2019 IEEE Global Communications Conference (GLOBECOM), pp. 1–5 (2019)
52. Ahmed, N., Hussain, M.I.: Periodic traffic scheduling for IEEE 802.11ah networks. IEEE Commun. Lett. **24**(7), 1510–1513 (2020)
53. Seferagić, A., Moerman, I., De Poorter, E., Hoebeke, J.: Evaluating the suitability of IEEE 802.11ah for low-latency time-critical control loops. IEEE Internet Things J. **6**(5), 7839–7848 (2019)
54. Kim, J., Lee, H.W., Chong, S.: Super-MAC design for tightly coupled multi-RAT networks. IEEE Trans. Commun. **67**(10), 6939–6951 (2019)
55. Ye, Q., Zhuang, W., Li, L., Vigneron, P.: Traffic-load-adaptive medium access control for fully connected mobile ad hoc networks. IEEE Trans. Veh. Technol. **65**(11), 9358–9371 (2016)
56. Shahin, N., Ali, R., Kim, Y.T.: Hybrid slotted-CSMA/CA-TDMA for efficient massive registration of IoT devices. IEEE Access **6**, 18,366–18,382 (2018)
57. Gao, J., Zhuang, W., Li, M., Shen, X., Li, X.: MAC for machine-type communications in industrial IoT–Part I: Protocol design and analysis. IEEE Internet Things J. **8**(12), 9945–9957 (2021)
58. Kim, D.M., Sorensen, R.B., Mahmood, K., Osterbo, O.N., Zanella, A., Popovski, P.: Data aggregation and packet bundling of uplink small packets for monitoring applications in LTE. IEEE Netw. **31**(6), 32–38 (2017)
59. Alonso, L., Agustí, R., Sallent, O.: A near-optimum MAC protocol based on the distributed queueing random access protocol (DQRAP) for a CDMA mobile communication system. IEEE J. Sel. Areas Commun. **18**(9), 1701–1718 (2000)
60. Dhillon, H.S., Huang, H., Viswanathan, H., Valenzuela, R.A.: Fundamentals of throughput maximization with random arrivals for M2M communications. IEEE Trans. Commun. **62**(11), 4094–4109 (2014)
61. Wang, P., Zhuang, W.: A collision-free MAC scheme for multimedia wireless mesh backbone. IEEE Trans. Wireless Commun. **8**(7), 3577–3589 (2009)
62. Malekshan, K.R., Zhuang, W., Lostanlen, Y.: An energy efficient MAC protocol for fully connected wireless ad hoc networks. IEEE Trans. Wireless Commun. **13**(10), 5729–5740 (2014)
63. Malekshan, K.R., Zhuang, W., Lostanlen, Y.: Coordination-based medium access control with space-reservation for wireless ad hoc networks. IEEE Trans. Wireless Commun. **15**(2), 1617–1628 (2015)

64. Gao, J., Li, M., Zhao, L., Shen, X.: Contention intensity based distributed coordination for V2V safety message broadcast. IEEE Trans. Veh. Technol. **67**(12), 12,288–12,301 (2018)
65. Jiang, D., Wang, H., Malkamaki, E., Tuomaala, E.: Principle and performance of semi-persistent scheduling for VoIP in LTE system. In: Proceedings of the 2007 International Conference on Wireless Communications, Networking, Mobile Computing, pp. 2861–2864 (2007)
66. Wang, P., Zhuang, W.: A token-based scheduling scheme for WLANs supporting voice/data traffic and its performance analysis. IEEE Trans. Wireless Commun. **7**(5), 1708–1718 (2008)
67. Gamage, A.T., Liang, H., Shen, X.: Two time-scale cross-layer scheduling for cellular/WLAN interworking. IEEE Trans. Commun. **62**(8), 2773–2789 (2014)
68. Ksentini, A., Frangoudis, P.A., Amogh, P., Nikaein, N.: Providing low latency guarantees for slicing-ready 5G systems via two-level MAC scheduling. IEEE Netw. **32**(6), 116–123 (2018)
69. Lioumpas, A.S., Alexiou, A.: Uplink scheduling for machine-to-machine communications in LTE-based cellular systems. In: Proceedings of the 2011 IEEE GLOBECOM Workshops (GC Wkshps), pp. 353–357 (2011)
70. Al-Janabi, T.A., Al-Raweshidy, H.S.: An energy efficient hybrid MAC protocol with dynamic sleep-based scheduling for high density IoT networks. IEEE Internet Things J. **6**(2), 2273–2287 (2019)
71. Si, P., Yang, J., Chen, S., Xi, H.: Adaptive massive access management for QoS guarantees in M2M communications. IEEE Trans. Veh. Technol. **64**(7), 3152–3166 (2015)
72. Karadag, G., Gul, R., Sadi, Y., Ergen, S.C.: QoS-constrained semi-persistent scheduling of machine-type communications in cellular networks. IEEE Trans. Wireless Commun. **18**(5), 2737–2750 (2019)
73. Zhang, C., Sun, X., Zhang, J., Wang, X., Jin, S., Zhu, H.: Throughput optimization with delay guarantee for massive random access of M2M communications in industrial IoT. IEEE Internet Things J. **6**(6), 10077–10092 (2019)
74. Arouk, O., Ksentini, A., Taleb, T.: Group paging-based energy saving for massive MTC accesses in LTE and beyond networks. IEEE J. Sel. Areas Commun. **34**(5), 1086–1102 (2016)
75. Salodkar, N., Karandikar, A., Borkar, V.S.: A stable online algorithm for energy-efficient multiuser scheduling. IEEE Trans. Mobile Comput. **9**(10), 1391–1406 (2010)
76. Chang, C.S., Lee, D.S., Wang, C.: Asynchronous grant-free uplink transmissions in multichannel wireless networks with heterogeneous QoS guarantees. IEEE/ACM Trans. Netw. **27**(4), 1584–1597 (2019)
77. Rodoplu, V., Nakıp, M., Eliiyi, D.T., Güzeliş, C.: A multiscale algorithm for joint forecasting–scheduling to solve the massive access problem of IoT. IEEE Internet Things J. **7**(9), 8572–8589 (2021)
78. Yang, B., Cao, X., Han, Z., Qian, L.: A machine learning enabled MAC framework for heterogeneous Internet-of-Things networks. IEEE Trans. Wireless Commun. **18**(7), 3697–3712 (2019)
79. Yang, B., Cao, X., Bassey, J., Li, X., Qian, L.: Computation offloading in multi-access edge computing: A multi-task learning approach. IEEE Trans. Mobile Comput. **20**, 2745–2762 (2020)
80. Kumcu, A., Bombeke, K., Platiša, L., Jovanov, L., Van Looy, J., Philips, W.: Performance of four subjective video quality assessment protocols and impact of different rating preprocessing and analysis methods. IEEE J. Sel. Topics Signal Process. **11**(1), 48–63 (2016)
81. Chollet, F., et al.: Keras: The python deep learning library. Astrophysics Source Code Library
82. Kingma, D.P., Ba, J.: Adam: A method for stochastic optimization (2014). arXiv:1412.6980
83. Srivastava, N., Hinton, G., Krizhevsky, A., Sutskever, I., Salakhutdinov, R.: Dropout: a simple way to prevent neural networks from overfitting. J. Mach. Learn. Res. **15**(1), 1929–1958 (2014)
84. Yoon, S., Li, L.E., Liew, S.C., Choudhury, R.R., Rhee, I., Tan, K.: Quicksense: fast and energy-efficient channel sensing for dynamic spectrum access networks. In: Proceedings of 2013 IEEE INFOCOM, pp. 2247–2255 (2013)

85. IEEE standard for information technology–telecommunications and information exchange between systems local and metropolitan area networks–specific requirements - part 11: Wireless LAN medium access control (MAC) and physical layer (PHY) specifications. IEEE Std 802.11-2016 (Revision of IEEE Std 802.11-2012) pp. 1–3534 (2016)
86. Gharbieh, M., ElSawy, H., Yang, H.C., Bader, A., Alouini, M.S.: Spatiotemporal model for uplink IoT traffic: Scheduling and random access paradox. IEEE Trans. Wireless Commun. **17**(12), 8357–8372 (2018)

Chapter 3
UAV-Assisted Edge Computing: Rural IoT Applications

In this chapter, we study unmanned aerial vehicle (UAV) assisted mobile edge computing to optimize computing offloading with minimum UAV energy consumption. In the considered scenario, a UAV plays the role of an aerial server to collect and process the computing tasks offloaded by ground devices. Given the service requirements of devices, we jointly optimize the UAV trajectory, the device transmit power, and computing load allocation. The resulting optimization problem corresponds to nonconvex fractional programming, and the Dinkelbach algorithm and the successive convex approximation technique are adopted to solve it. Furthermore, we decompose the problem into multiple subproblems for distributed and parallel problem-solving. Simulation results demonstrate the effectiveness of the proposed approach for maximizing the energy efficiency of the UAV.

3.1 Background on UAV-Assisted Edge Computing

MEC is a key enabling technology to support computing services for billions of IoT devices [1, 2]. IoT devices can offload their computing tasks to network edges to prolong their battery life and reduce computing delay, which benefits energy-constrained IoT devices, e.g., energy-harvesting devices [3], and devices with limited computing capability, e.g., smart cameras [4]. However, many IoT devices operate in unattended areas, such as forests, deserts, mountains, or underwater locations [5], to execute some compute-intensive applications, including pipeline monitoring and control [6], underwater infrastructure monitoring [7], and military operations [8]. In these scenarios, the IoT devices alone cannot support the applications, and the computing loads cannot be offload to the cloud server due to the sparsely deployed terrestrial communication infrastructures.

© The Author(s), under exclusive license to Springer Nature Switzerland AG 2021
J. Gao et al., *Connectivity and Edge Computing in IoT: Customized Designs and AI-based Solutions*, Wireless Networks,
https://doi.org/10.1007/978-3-030-88743-8_3

The advancement of network infrastructure and communication technologies facilitates a new solution, i.e., using UAVs to support the connectivity of IoT devices in rural areas. UAVs equipped with communication and computing capability can serve rural IoT devices. As a key technology in next-generation networks, UAV-assisted networks have drawn significant attention from academia in recent years for applications related to ubiquitous communication [9, 10], low latency and real-time communication [11, 12], and post-disaster recovery [13, 14]. Extensive performance tests and communication protocol development on UAV-assisted networks have been conducted in recent years by 3GPP [15], which demonstrates the ability of UAVs to provide flexible and ubiquitous services for IoT devices. With computing capabilities, UAVs can function as mobile edge servers to collect and process computing tasks of ground IoT devices that cannot connect to terrestrial edge servers. The potential advantages of UAV-assisted computing networks can be summarized as follows:

(1) Flexible computing capability deployment—UAVs can be dispatched to designated places for providing on-demand communication and computing services to IoT devices [9, 16–18];
(2) Reliable offloading links—UAVs operate at a high altitude, providing line-of-sight (LoS) communication links to ground IoT devices. The LoS links facilitate reliable computing offloading;
(3) Cost-effectiveness—Compared with deploying fixed terrestrial infrastructures, UAV-mounted servers offer a low-cost solution to deal with the computing demands that are spatio-temporally changing.

In comparison with terrestrial network infrastructures, UAVs have their own features, which result in additional design requirements for UAV-assisted computing networks. The features of UAVs include

(1) Limited computing energy—The on-board energy of UAVs is limited. Thus, UAV-mounted servers cannot provide durable computing services to IoT devices;
(2) Maneuverability—While the maneuverability of UAVs increases the flexibility in communication service deployment, it yields dynamic channel conditions and operation constraints for computing resource deployment.

Moreover, the maneuverability of UAVs increases the energy demand due to the energy consumption from aircraft engines. A UAV-mounted server needs to move to collect data from sparsely distributed devices, while a significant portion of UAV energy consumption is for flying around to collect data. Therefore, an energy-efficient UAV trajectory and computing offloading strategy is important for UAV-assisted MEC.

3.2 Connectivity Requirements of UAV-Assisted MEC for Rural IoT

UAV-assisted MEC provides a connectivity solution for IoT devices in rural areas. Meanwhile, UAV-assisted networks and rural IoT introduce requirements and constraints in connectivity.

3.2.1 Network Constraints

As aforementioned, the on-board energy of a UAV is limited. Consider the two popular types of UAVs in the market: fixed-wing UAVs and rotary-wing UAVs. The battery power of quad-rotor and fixed-wing UAVs can only support up to an hour and a few days of service, respectively [19]. The maneuverability of quad-rotor UAVs is relatively higher than that of fixed-wing UAVs. Thus, the most common application for quad-rotor UAVs is to provide static coverage while hovering in a designated area. On the other hand, fixed-wing UAVs have high horizontal speed but cannot hover. Thus, the most common application for fixed-wing UAVs is to collect computing tasks by flying above the ground devices following a designed trajectory. A comparison between fixed-wing and rotary-wing UAVs is provided in Table 3.1. The trajectory of UAVs should be properly designed, considering the features of UAVs.

IoT devices in rural areas usually have no access to power grids. Therefore, they may prefer to offload as many computing tasks to an edge server as possible, in order to reduce their energy consumption on computing. Thus, it is necessary for a UAV to fly around for collecting data from sparsely distributed devices, which consumes energy. As a result, the UAV trajectory should be carefully determined to balance the number of computing tasks collected by the UAV and the energy consumed for collecting tasks, while finding the optimal trajectory is nontrivial.

Last, the computing energy consumption of a UAV may not be negligible even though it is relatively small compared to the mechanical energy consumption of the UAV. In the state-of-the-art MEC server architecture, dynamic voltage and frequency scaling (DVFS) is adopted to adjust the power setting on a computing device's processors and maximize power saving in computing. With DVFS, the

Table 3.1 Comparison of between fixed-wing and rotary-wing UAVs

	Maximum height	Horizontal speed	Vertical speed	Hovering	Endurance
Fixed-wing	16 km	Fast	Medium	Not supported	From half a day to days or weeks
Rotary-wing	6 km	Medium	Fast	Supported	Less than 1 h (on LiPo battery)

computing energy consumption for a unit time grows cubically as the computing load increases [20]. Without proper computing load allocation, the computing energy consumption can be excessive. Another possible consequence is that the offloaded tasks cannot be finished in time. Furthermore, the computing load allocation depends on the number of computing tasks offloaded by IoT devices and collected by the UAV. Therefore, UAV trajectory design, computing load allocation, and communication resource management are correlated in UAV-assisted MEC [21], which makes related designs very challenging.

3.2.2   State-of-the-Art Solutions

UAV-assisted networks have been investigated in [22–24]. In [22], Wu et al. consider trajectory design and downlink communication power control for a multi-UAV multi-device system to maximize the throughput of ground devices in a downlink scenario. In [23], Zeng et al. analyze the energy efficiency of a UAV-assisted network and design a UAV trajectory for hovering above a single ground mobile device. In [24], Tang et al. investigate a game-based channel assignment scheme for UAVs in D2D-enabled communication networks. UAVs have also been utilized to enhance the flexibility of MEC in [25, 26], where UAVs act as communication relays to assist the computing offloading of ground devices. Recently, more works utilize UAVs as aerial servers to provide edge computing services [27–29]. In [27], Jeong et al. study UAV trajectory planning to minimize communication energy consumption for offloading tasks from mobile devices, given a limited energy budget of the UAV-mounted server for computing. In [28], Tang et al. propose a UAV-assisted recommendation system in location-based social networks, while a UAV-mounted server is deployed to reduce computing and traffic load at a cloud server. In [29], Cheng et al. propose a computing offloading strategy in an IoT network, given pre-determined UAV trajectories. The work aims to jointly minimize the computing delay, device energy consumption, and server computing cost, although the energy consumption of the UAV-mounted server is not investigated. None of the above works discusses the energy efficiency for computing in a UAV-mounted server, which is important for prolonging the computing service lifetime of the UAV.

3.3   Multi-Resource Allocation for UAV-Assisted Edge Computing

To address the connectivity requirements, we investigate energy-efficient resource management for UAV-assisted MEC in this chapter. IoT devices on the ground can access and partially offload their computing tasks to a UAV-mounted server according to their service requirements. The UAV flies according to a designated trajectory to collect the offloading data, process computing tasks, and send computing results back to the devices. We aim to optimize the energy efficiency of the UAV, which is defined as the ratio of the overall offloaded computing data to the UAV energy consumption, by jointly optimizing the UAV trajectory and resource allocation in communication and computing. The highlights of this chapter are summarized as follows:

(1) A model for energy-efficient resource allocation for a UAV-assisted MEC system is developed. Based on the model, communication and computing resources are allocated, subject to device communication energy budgets, computing capability, and the mechanical operation constraints of the UAV;
(2) We exploit optimization techniques to solve the non-convex resource allocation problem. In order to improve scalability, we further adopt the alternating direction method of multipliers (ADMM) technique to facilitate distributed optimization;
(3) A spatial distribution estimation technique, i.e., Gaussian kernel density estimation, is applied to predict the location of ground devices. Based on the predicted location information, our proposed strategy can determine an energy-efficient UAV trajectory when the device mobility and offloading requests are ambiguous.

3.3.1 Network Model

A UAV-assisted MEC system is shown in Fig. 3.1, in which a single UAV-mounted server is deployed to offer edge computing services for ground devices in area $\mathcal{A}$. The UAV periodically collects and processes the computing tasks offloaded from ground devices. Each device processes the rest of the computing tasks locally if the task cannot be fully collected by the UAV. Define the computing cycle as a duration of T seconds. Each cycle contains K discrete time slots with equal length. Denote the set of time slots in the cycle by $\mathcal{K}$. Thus, the time length for a slot is T/K, which is denoted by Δ.

At the beginning of each cycle, ground devices with computing tasks in area $\mathcal{A}$ send offloading requests to the UAV-mounted server. Denote the set of those ground devices by $\mathcal{I}$, where $\mathcal{I} = \{1, \ldots, N\}$. Assume the ground devices in $\mathcal{I}$ can connect to the UAV for all time slots in a cycle. In this work, the UAV and the devices cooperatively determine the offloading and resource allocation strategy for each cycle, including the UAV moving trajectory, the transmit power of ground devices, and computing load allocation for the UAV-mounted server.[1] During a cycle, the UAV flies over the ground devices and offers the computing service according to

---

[1] We assume that the computing load of finding the optimal strategy is negligible as compared with the computing loads of the offloaded tasks.

Fig. 3.1 System model

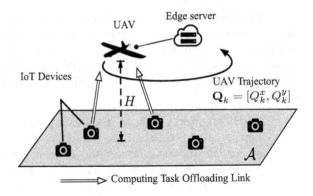

➡ Computing Task Offloading Link

the designated trajectory and resource allocation strategy. By the end of the cycle, the UAV returns to a predetermined final position.

3.3.2   Communication Model

The quality of communication links between the UAV and ground devices depends on their locations. To represent their locations, we construct a 3D Cartesian coordinate system. For IoT device i, the horizontal coordinate at time k is denoted by $\mathbf{q}_{i,k} = [q_{i,k}^x, q_{i,k}^y]$. Assume that devices know their trajectory for the upcoming cycle, i.e., $\{\mathbf{q}_{i,k}, \forall k\}$. For the UAV, the horizontal coordinate at time k is denoted by $\mathbf{Q}_k = [Q_k^x, Q_k^y]$. The UAV moves at a fixed altitude, H. The UAV trajectory plan, as an optimization variable, consists of UAV positions in the whole cycle, i.e., $\mathbf{Q} = [\mathbf{Q}_1; \ldots; \mathbf{Q}_K]$. The average UAV velocity in slot k is given by

$$\mathbf{v}_k(\mathbf{Q}) = \frac{\mathbf{Q}_k - \mathbf{Q}_{k-1}}{\Delta}, \forall k \geq 2. \tag{3.1}$$

The average acceleration in slot k is given by

$$\mathbf{a}_k(\mathbf{Q}) = \frac{\mathbf{v}_k(\mathbf{Q}) - \mathbf{v}_{k-1}(\mathbf{Q})}{\Delta}, \forall k \geq 2. \tag{3.2}$$

The magnitudes of velocity and acceleration are constrained by the maximum speed and the maximum acceleration, which are denoted by $v_{\max}$ and $a_{\max}$, respectively.

It is assumed that the Doppler frequency shift in the communication can be compensated at the receiver. The channel quality depends on the distance between the UAV and devices. Due to the high probability of LoS links in UAV communication [23], we assume that the channel gain follows a free-space path loss model. The channel gain for device i in slot k is denoted by $h_{i,k}$, given by

$$h_{i,k}(\mathbf{Q}_k) = \frac{g_0}{\|\mathbf{Q}_k - \mathbf{q}_{i,k}\|_2^2 + H^2} \tag{3.3}$$

where $\|\cdot\|_2$ is L2 norm, and g_0 denotes the received power at the reference distance (e.g., $d = 1$ m) between the transmitter and the receiver. We consider two channel access schemes: (i) orthogonal access, in which the radio spectrum bandwidth is partitioned into N sub-channels each occupied by one device; and (ii) non-orthogonal access, in which the bandwidth is shared among devices. Denote the channel bandwidth for the uplink by B. The amount of data that can be offloaded by device i in slot k is

$$R_{i,k}(\delta_{i,k}, \mathbf{Q}_k) = \frac{B\Delta}{N} \log\left[1 + \frac{\delta_{i,k}h_{i,k}(\mathbf{Q}_k)P}{\sigma^2(B/N)}\right] \tag{3.4}$$

under the orthogonal access model, and

$$R_{i,k}(\delta_k, \mathbf{Q}_k) = B\Delta\log\left[1 + \frac{\delta_{i,k}h_{i,k}(\mathbf{Q}_k)P}{\sigma^2 B + \sum_{j\neq i}\delta_{j,k}h_{j,k}(\mathbf{Q}_k)P}\right] \tag{3.5}$$

under the non-orthogonal channel model. In (3.4) and (3.5), P and σ^2 denote the maximum transmit power of ground devices and the power spectral density of channel noise, respectively; $\delta_{i,k} \in [0, 1]$ represents the portion of the maximum power that is allocated to device i within time slot k, which is a part of the offloading strategy; δ_k denotes the vector of $\delta_{i,k}$ for all $i \in \mathcal{I}$ in slot k. The received noise power in the transmission is represented by n_0, where $n_0 = \sigma^2 B/N$ for the orthogonal channel access model, and $n_0 = \sigma^2 B$ for the non-orthogonal channel access model. In non-orthogonal model, devices share the same channel to offload their tasks. The communication of a device may interfere with that of other devices.

3.3.3 Computing Model

Due to the limited battery and the computing capability of the UAV, only a part of tasks can be offloaded and executed in the UAV-mounted server. Full granularity in task partition is considered, where the task-input data can be arbitrarily divided for local and remote executions [27, 30, 31]. Accordingly, a portion of the computing tasks are offloaded to the server while the rest are executed by the ground devices locally. Devices upload the input data for their offloaded tasks, and the UAV processes the corresponding computing loads of those tasks. We assume that the computing load can be executed once the input data is received, and the amount of data to process is equal to the size of input bits of tasks [27]. A task partition technique is considered, which divides the input bits between the offloaded computing load and local computing load. The overall input data size for computing tasks of device i is denoted by I_i. We set a threshold $\check{I}_i$ as the minimum the amount of input data required to be offloaded to the server for device i, where $\check{I}_i \leq I_i$. The threshold represents the part of computing tasks to be processed in the server. Thus, the overall offloaded bits of device i is constrained as follows:

$$\check{I}_i \leq \sum_{k \in \mathcal{K}} R_{i,k}(\boldsymbol{\delta}_k, \mathbf{Q}_k) \leq I_i, \forall i. \tag{3.6}$$

In a scenario that (3.6) holds, if tasks cannot be fully offloaded, the rest of the tasks are processed by IoT devices locally.

After devices upload the input data, the UAV will store the received data in a buffer for further processing. The UAV processes the received data according to the workload allocation results. Let $W_{i,k}$ denote the amount of data from device i's offloaded task, to be processed in slot k. The UAV can only compute the task which is offloaded and received, and all offloaded tasks should be executed by the end of the cycle. Therefore, the following computing constraints are given:

$$\sum_{t=1}^{k} R_{i,t}(\boldsymbol{\delta}_k, \mathbf{Q}_k) \geq \sum_{t=1}^{k} W_{i,t}, \forall k \tag{3.7a}$$

$$\sum_{t=1}^{K} R_{i,t}(\boldsymbol{\delta}_k, \mathbf{Q}_k) = \sum_{t=1}^{K} W_{i,t}. \tag{3.7b}$$

In addition, for local computing, the CPU-cycle frequency of the IoT device i is a constant, denoted by f_i^M. For the UAV-mounted server, we consider a CPU featuring the DVFS technique. The CPU-cycle frequency can step up or down according to the computing workload and is bounded by the maximum CPU-cycle frequency, f_{max}^U. As in [20, 30], the CPU-cycle frequency for the server can be calculated by

$$f_k^U(\mathbf{W}_k) = \frac{\sum_i \chi_i W_{i,k}}{\Delta} \leq f_{max}^U, \forall k \tag{3.8}$$

where $f_k^U(\mathbf{W}_k)$ represents the CPU-cycle frequency in time slot k, and χ_i denotes the number of computing cycles needed to process 1 bit of data.

3.3.4 Energy Consumption Model

Energy Consumption at devices The main energy consumption of devices is the energy cost from communication and local computing. Firstly, the communication energy consumption of device i for offloading tasks in slot k can be formulated as

$$S_{i,k}(\delta_{i,k}) = \delta_{i,k} P \Delta. \tag{3.9}$$

The overall offloading communication energy of device i is bounded by E_i^T, i.e.,

$$\sum_k S_{i,k}(\delta_{i,k}) \leq E_i^T, \forall i. \tag{3.10}$$

Therefore, the energy consumption of a device on communication can be reduced if the UAV is closer. On the other hand, for the computing energy consumption, we consider a lower bound of offloaded bits $\check{I}_i$, which ensures that the remaining computing load at the device will not consume excessive computing energy locally, i.e.,

$$E_i^M = \kappa \chi_i (I_i - \check{I}_i)(f_i^M)^2 \leq \hat{E}_i^M \tag{3.11}$$

where E_i^M is the maximum computing energy that can be reached by threshold $\check{I}_i$, and $\hat{E}_i^M$ is a parameter representing the constraint of computing energy consumption. The computing energy model is adopted from [20, 32]. Parameters f_i^M and κ represent the fixed CPU-cycle frequency of device i and a constant related to the hardware architecture, respectively.

Energy Consumption at UAV-Mounted Server The main energy consumption at the UAV-mounted server consists of the energy cost from mechanical operation and computing. Although downlink transmission is needed in our system, this part of energy consumption is negligible for two reasons: (1) The communication energy is small compared to the UAV propulsion and computing energy; (2) The output computing results usually have much less data amount compared to the input data amount [33]. We use the refined UAV propulsion energy consumption model for fixed-wing UAV following [23].[2] The propulsion energy consumption in slot k depends on the instantaneous UAV acceleration and velocity, given by

$$E_k^F(\mathbf{Q}) = \gamma_1 \|\mathbf{v}_k(\mathbf{Q})\|_2^3 + \frac{\gamma_2}{\|\mathbf{v}_k(\mathbf{Q})\|_2}\left(1 + \frac{\|a_k(\mathbf{Q})\|_2^2}{g^2}\right) \tag{3.12}$$

where g denotes the gravitational acceleration; γ_1 and γ_2 are fixed parameters related to the aircraft's weight, wing area, air density, and so on [23, 27]. The computing energy for executing tasks from device i in time slot k is expressed as

$$E_{i,k}^{C,U}(\mathbf{W}_k) = \kappa \chi_i W_{i,k}\left(f_k^U(\mathbf{W}_k)\right)^2. \tag{3.13}$$

---

[2] We deploy the fixed-wing UAV in the proposed system as an example. The proposed approach can be extended to the system with a quad-rotor UAV, where the mechanical energy consumption model is different.

3.3.5 Problem Formulation

Our main objective is to maximize the energy efficiency of the UAV-mounted server, subject to device task offloading constraints, UAV computing capabilities, and the mechanical constraints of the UAV. The energy efficiency of the UAV is defined as the ratio of the overall offloaded data to the energy consumption of the UAV in a cycle. The energy efficiency maximization problem is formulated as follows.

$$\max_{\delta, \mathbf{W}, \mathbf{Q}} \quad \eta = \frac{\sum_{i \in \mathcal{I}} \sum_{k \in \mathcal{K}} R_{i,k}(\delta_k, \mathbf{Q}_k)}{\sum_{k \in \mathcal{K}} \sum_{i \in \mathcal{I}} E_{i,k}^{C,U}(\mathbf{W}_k) + \sum_{k \in \mathcal{K}} E_k^F(\mathbf{Q})} \tag{3.14a}$$

$$\text{s.t.} \quad \|\mathbf{v}_k(\mathbf{Q})\|_2 \le v_{max}, \forall k, \tag{3.14b}$$

$$\|\mathbf{a}_k(\mathbf{Q})\|_2 \le a_{max}, \forall k, \tag{3.14c}$$

$$\mathbf{Q}_K = \mathbf{Q}_f, \mathbf{v}_K(\mathbf{Q}) = \mathbf{v}_0, \tag{3.14d}$$

$$0 \le \delta_{i,k} \le 1, \tag{3.14e}$$

$$(3.6), (3.7a), (3.7b), (3.8), (3.10).$$

In (3.14), $\mathbf{Q}_f$ represents the designated final position of the UAV, and $\mathbf{v}_0$ represents the initial velocity at the beginning of the cycle. The constraints can be categorized into three types: 1) device QoS constraints (3.6), (3.10), and (3.14e); 2) UAV computing capacity constraints (3.7a), (3.7b), and (3.8); 3) UAV mechanical constraints (3.14b), (3.14c), and (3.14d). The optimization problem is a non-linear fractional programming. Due to the interference among devices in the non-orthogonal channel and the propulsion energy consumption for the fixed-wing UAV, both functions $R_{i,k}(\delta_k, \mathbf{Q}_k)$ and $E_k^F(\mathbf{Q})$ are non-convex. Therefore, solving optimization problem (3.14) is challenging. Finding the optimal solution of a non-convex problem is often slow and may not be possible. In the following, we propose an approach to find a local optimum efficiently.

3.3.6 Optimization Algorithm for UAV-Assisted Edge Computing

In this section, an optimization approach is introduced to find a solution of problem (3.14). Firstly, an inner convex approximation method is applied to approximate the non-convex functions, $R_{i,k}(\delta_k, \mathbf{Q}_k)$, and $E_k^F(\mathbf{Q})$ by solvable convex functions. The successive convex approximation (SCA) based algorithm is adopted to achieve the local optimum of the original problem. After the approximated convex functions are built, the fraction programming in the inner loop of the SCA-based algorithm is handled by the Dinkelbach algorithm. Moreover, in order to improve scalability, the problem is further decomposed into several sub-problems via the ADMM technique, in which the power allocation is solved by devices in a distributed manner, while the computing load allocation and UAV trajectory planning are determined by the UAV.

Problem (3.14) is a non-convex problem due to $R_{i,k}(\delta_k, \mathbf{Q}_k)$ and $E_k^F(\mathbf{Q})$. To construct an approximation that is solvable, we first introduce several auxiliary variables, $\{\xi_{i,k}, \omega_k, l_{i,k}, A_k, \check{R}_{i,k}, \hat{E}_{i,k}^F\}$. For the orthogonal channel access scheme, the equivalent form of problem (3.14) is as follows:

$$\max_{\mathcal{V}} \quad \check{\eta}(\mathcal{V}) = \frac{\sum_{i\in\mathcal{I}} \sum_{k\in\mathcal{K}} \check{R}_{i,k}}{\sum_{k\in\mathcal{K}} \sum_{i\in\mathcal{I}} E_{i,k}^{C,U}(\mathbf{W}_k) + \sum_{k\in\mathcal{K}} \hat{E}_k^F} \tag{3.15a}$$

$$\text{s.t.} \quad \check{R}_{i,k} \leq \frac{B\Delta}{N} \log(1 + \xi_{i,k}), \forall i, k \tag{3.15b}$$

$$\xi_{i,k} l_{i,k} \leq \delta_{i,k} P, \forall i, k \tag{3.15c}$$

$$\frac{(\|\mathbf{Q}_k - \mathbf{q}_{i,k}\|_2^2 + H^2) n_0}{g_0} \leq l_{i,k}, \forall i, k \tag{3.15d}$$

$$\hat{E}_k^F \geq \gamma_1 \|\mathbf{v}_k(\mathbf{Q})\|_2^3 + \gamma_2 A_k, \forall k \tag{3.15e}$$

$$\omega_k^2 \leq \|\mathbf{v}_k(\mathbf{Q})\|_2^2, \forall k \tag{3.15f}$$

$$\omega_k A_k \geq 1 + \frac{\|a_k(\mathbf{Q})\|_2^2}{g^2}, \forall k \tag{3.15g}$$

$$\check{I}_i \leq \sum_{k\in\mathcal{K}} \check{R}_{i,k} \leq I_i, \forall i, \tag{3.15h}$$

$$(3.6), (3.7a), (3.7b), (3.8), (3.10), (3.14b)–(3.14e).$$

Set $\mathcal{V}$ represents the union set of the primary and auxiliary optimization variables, where $\mathcal{V} = \{\delta, \mathbf{W}, \mathbf{Q}, \xi, \omega, \mathbf{l}, \mathbf{A}, \check{\mathbf{R}}, \hat{\mathbf{E}}^F\}$. For the non-orthogonal channel model, constraint (3.15b) is replaced by the following constraint:

$$\check{R}_{i,k} \leq B\Delta \Big[\log\big(1 + \sum_{i\in\mathcal{I}} \xi_{i,k}\big) - \log\big(1 + \sum_{j\in\mathcal{I}/\{i\}} \xi_{j,k}\big) \Big], \forall i, k. \tag{3.16}$$

Lemma 3.1 *Problem (3.15) is an equivalent form of problem (3.14).*

Proof Firstly, to deal with the non-convex function on the numerator, i.e., $R_{i,k}(\delta_{i,k}, \mathbf{Q}_k)$, we introduce auxiliary variable $\check{R}_{i,k}$ to indicate the lower bound of the data rate for device i in slot k. Moreover, we introduce two auxiliary variables: $\xi_{i,k}$, where $\xi_{i,k} \leq \delta_{i,k} P / l_{i,k}$, and $l_{i,k}$, where $l_{i,k} \geq N_0 / h_{i,k}$. Thus, the following relation can be established

$$\check{R}_{i,k} \leq \frac{B\Delta}{N} \log(1 + \xi_{i,k}) \leq R_{i,k}(\delta_{i,k}, \mathbf{Q}_k) \tag{3.17}$$

where $\check{R}_{i,k}$ is the epigraph form of $R_{i,k}(\delta_{i,k}, \mathbf{Q}_k)$. When (3.15a) is maximized, *i.e.*, the numerator $\check{R}^*_{i,k}$ is maximized, we have $l^*_{i,k} = 1/g^*_{i,k}$, $\xi^*_{i,k} = \delta^*_{i,k}P/l^*_{i,k}$, and $\check{R}^*_{i,k} = R_{i,k}(\delta^*_{i,k}, \mathbf{Q}^*_k)$.

Furthermore, to deal with the non-linear function in the denominator, *i.e.*, $E^F_k(\mathbf{Q})$, we introduce auxiliary variable $\hat{E}^F_k$ to indicate the upper bound of the UAV propulsion energy in slot k. For the non-linear part of the function, we introduce two auxiliary variables: ω_k, where $\omega^2_k \leq \|\mathbf{v}_k(\mathbf{Q})\|_2^2$, and $A_{i,k}$, where $A_{i,k} \geq (1/\omega_k)(1 + \|a_k(\mathbf{Q})\|_2^2/g^2)$. Thus, we have

$$
\hat{E}^F_k \geq \gamma_1 \|\mathbf{v}_k(\mathbf{Q})\|_2^3 + \gamma_2 A_k
$$

$$
\geq \gamma_1 \|\mathbf{v}_k(\mathbf{Q})\|_2^3 + \gamma_2 \frac{1}{\omega_k}(1 + \frac{\|a_k(\mathbf{Q})\|_2^2}{g^2}) \geq E^F_k(\mathbf{Q}). \tag{3.18}
$$

Similarly, when (3.15a) is maximized, *i.e.*, the denominator, $E^F_k(\mathbf{Q})$, is minimized, $\hat{E}^{F*}_k = E^F_k(\mathbf{Q}^*)$. Therefore, problem (3.15) is equivalent to problem (3.14), and $\eta^* = \check{\eta}^*$. $\qquad\square$

Problem (3.15) includes four non-convex constraints, which are (3.15c), (3.15f), (3.15g), and (3.16). We approximate those non-convex constraints by their first order Taylor expansions and adopt the successive convex optimization technique to solve the problem. New auxiliary variables, $\{\xi^t_{i,k}, l^t_{i,k}, \omega^t_k, A^t_k, \mathbf{v}_k, z^t_{i,k}\}$, are introduced to represent the corresponding estimated optimizers at the previous iteration of optimization, i.e., iteration t. The SCA-based algorithm iterates until the estimated solution reaches to a local optimum. Constraint (3.15c) can be approximated as follows:

$$
\|\xi_{i,k} + l_{i,k}, \xi^t_{i,k} - l^t_{i,k}, x_{i,k} - 1\|_2 \leq x_{i,k} + 1 \tag{3.19}
$$

where

$$
x_{i,k} = \delta_{i,k}P - \frac{(\xi^t_{i,k} - l^t_{i,k})(\xi_{i,k} - l_{i,k})}{2}.
$$

Constraint (3.15f) can be approximated by

$$
\omega^2_k \leq \|\mathbf{v}^t_k\|_2^2 + 2(\mathbf{v}^t_k)^T(\mathbf{v}_k(\mathbf{Q}) - \mathbf{v}^t_k). \tag{3.20}
$$

Constraint (3.15g) can be approximated by

$$
\|\omega_k - A_k, \omega^t_k + A^t_k, y_k - 1, 2, \frac{2a_k(\mathbf{Q}_k)}{g}\|_2 \leq y_k + 1 \tag{3.21}
$$

where

$$
y_k = \frac{(\omega^t_k + A^t_k)(\omega_k + A_k)}{2}.
$$

Constraint (3.16) can be approximated as follows:

$$\check{R}_{i,k} \leq \frac{B\Delta}{N} \Big[\log(1 + \xi_{i,k} + e_{i,k}) - \log(1 + e_{i,k}^t) - \frac{e_{i,k} - e_{i,k}^t}{\ln 2(1 + e_{i,k}^t)} \Big] \tag{3.22}$$

where $e_{i,k} = \sum_{j \in \mathcal{I}/\{i\}} \xi_{i,k}$.

Lemma 3.2 *Non-convex constraints (3.15c), (3.15f), (3.15g), and (3.16) can be approximated by the convex forms in (3.19)–(3.22). The solution of the approximated problem is a local maximizer of problem (3.14), which provides the lower bound of the maximum energy efficiency that can be achieved.*

Proof Constraint (3.15c) can be transformed into the following equivalent form:

$$(\xi_{i,k} + l_{i,k})^2 - (\xi_{i,k} - l_{i,k})^2 \leq 4\delta_{i,k} P \tag{3.23}$$

which is difference of convex functions [34]. Then, we approximate the second part of the equation by Taylor expansion,

$$(\xi_{i,k} - l_{i,k})^2 \approx (\xi_{i,k}^t - l_{i,k}^t)^2 + \begin{bmatrix} 2\xi_{i,k}^t - 2l_{i,k}^t \\ 2l_{i,k}^t - 2\xi_{i,k}^t \end{bmatrix}^T \begin{bmatrix} \xi_{i,k} - \xi_{i,k}^t \\ l_{i,k} - l_{i,k}^t \end{bmatrix} \tag{3.24}$$

Further, we reformulate the approximated equation as the constraints in (3.19) with a cone expression. Moreover, constraint (3.15g) is approximated by constraint (3.21) in a similar way. Constraints (3.15f) and (3.16) are approximated by (3.20) and (3.22), respectively, by first order Taylor expansion to obtain the lower bound on the squared norm and the subtracted term.

All the approximated constraints, (3.19)–(3.22), are stricter than their original counterparts, guaranteeing that the solution of the approximated problem is strictly smaller than the original optimum. For example, consider the optimal $\xi_{i,k}$ and $l_{i,k}$ obtained by solving the approximated problem, which are denoted by $\xi_{i,k}^a$ and $l_{i,k}^a$. These two variables are bounded by constraint (3.19) in the approximated problem. Comparing (3.19) with the original constraint (3.15c) and considering the property of the Taylor expansion, we have $\xi_{i,k}^a l_{i,k}^a + \Delta_{approx} \leq \delta_{i,k} P$, where $\Delta_{approx} \geq 0$. Thus,

$$\frac{B\Delta}{N} \log(1 + \xi_{i,k}^a) \leq \frac{B\Delta}{N} \log(1 + \frac{\delta_{i,k} P}{l_{i,k}^a}). \tag{3.25}$$

Moreover, due to $l_{i,k}^a \geq 1/g_{i,k}$, we have

$$\frac{B\Delta}{N} \log(1 + \frac{\delta_{i,k} P}{l_{i,k}^a}) \leq R_{i,k}(\delta_{i,k}, \mathbf{Q}_k). \tag{3.26}$$

Therefore, the approximation on constraint (3.15c) leads to $\check{R}^*_{i,k} < R_{i,k}(\delta_{i,k}, \mathbf{Q}_k)$. Other approximated constraints can be proven similarly to show that the approximated objective function provides the global lower bound for original objective function (3.14). Moreover, due to the gradient consistency in the first order estimation, the SCA algorithm will stop when a local optimizer is found.   □

Based on Lemmas 3.1 and 3.2, the SCA-based algorithm is summarized in Algorithm 3, where $\check{\eta}(\mathcal{V}; \mathbb{A}^t)$ represents the energy efficiency $\check{\eta}(\mathcal{V})$ in (3.15) with the given value in auxiliary variable set $\mathbb{A}^t$. Note that the approximated problem inside the loop (Steps 3 and 4 in Algorithm 3) is a fractional programming problem and non-convex. We will find the optimal solution of the approximated problem in the remainder of the section. The convergence of SCA has been proven in [34], and the algorithm will stop after a finite number of iterations if a local optimum exists.

Algorithm 3 SCA-based algorithm for solving problem (3.15)

1: Initialize the auxiliary variables $\mathbb{A}^0 = \{\xi^0_{i,k}, \omega^0_k, l^0_{i,k}, A^0_k, \check{R}^0_{i,k}, \hat{E}^{F,0}_{i,k}\}$ and loop index $t = 0$.
2: Solve the approximated problem (3.27) for given $\mathbb{A}^t$, and denote the optimal solution for auxiliary variables by $\mathbb{A}^{t+1}$:

$$\max_{\mathcal{V}} \qquad \check{\eta}(\mathcal{V}; \mathbb{A}^t) \qquad\qquad\qquad (3.27)$$

$$\text{s.t.} \qquad (3.6), (3.7a), (3.7b), (3.8), (3.10), (3.14b)\text{–}(3.14e),$$

$$(3.15d), (3.15e), (3.15h), (3.19)\text{–}(3.20),$$

$$(3.15b), \text{ in the case of orthogonal channel,}$$

$$(3.22), \text{ in the case of non-orthogonal channel.}$$

3: Update $t = t+1$. The difference of the solutions between two adjacent iterations, i.e., $\|\mathbb{A}^{t+1} - \mathbb{A}^t\|$, is below a threshold θ_1.

Problem (3.27) is a fraction programming problem. We can adopt the Dinkelbach algorithm to find the optimal solution. The objective function, (3.27), can be rewritten in the following parametric programming form,

$$F^t(\alpha) = \max_{\mathcal{V}} \left\{ \sum_{k\in\mathcal{K}}\sum_{i\in\mathcal{I}} \check{R}_{i,k} - \alpha\Big[\sum_{k\in\mathcal{K}}\sum_{i\in\mathcal{I}} E^{C,U}_{i,k}(\mathbf{W}_k) \right.$$
$$\left. + \sum_{k\in\mathcal{K}} \hat{E}^F_k \Big] | \mathcal{V} \in \mathcal{F}^t \right\} \qquad (3.28)$$

where $\mathcal{F}^t$ represents the feasible set of problem (3.27) at the t-th iteration in Algorithm 3. The function, $F^t(\alpha)$, is a monotonic decreasing function of α. Let α^* denote the solution of $F^t(\alpha^*) = 0$. Due to the monotone decreasing property of

---

Algorithm 4 Dinkelbach algorithm for solving problem (3.27)

1: Initialize $\alpha^0 = 0$ if $t = 0$, $\alpha^0 = \alpha^*$ in loop $t - 1$ if $t \geq 0$, and the loop index $m = 0$.
2: Solve problem (3.28) for given α^m, and denote the solution for the problem by $\mathcal{V}_d^m$.
3: Update the Dinkelbach auxiliary variable $\alpha^{m+1} = \check{\eta}(\mathcal{V}_d^m; \mathbb{A}^t)$.
4: $m = m + 1$. $F^t(\alpha^{m+1}) \leq \theta_2$.

$F^t(\alpha)$, $F^t(\alpha^*) = 0$ if and only if α^* is equal to the optimal result of problem (3.27), i.e., $\alpha^* = \check{\eta}(\mathcal{V}^*; \mathbb{A}^t)$ [35]. The algorithm for solving problem (3.27) is shown in Algorithm 4.

Due to the nature of the SCA-based algorithm and Dinkelbach algorithm, we can further reduce the number of iterations based on the following Lemma.

Lemma 3.3 *Denote optimal Dinkelbach parameter α^* for two consecutive SCA iterations by $\alpha^*(t - 1)$ and $\alpha^*(t)$. We have $\alpha^*(t - 1) \leq \alpha^*(t)$, and $F^t(\alpha^*(t - 1)) \geq F^t(\alpha^*(t)) = 0$.*

Proof Denote the optimization result and the corresponding Dinkelbach parameter at iteration $t - 1$ by $\mathcal{V}^*(t - 1)$ and $\alpha^*(t - 1)$, respectively. From Dinkelbach algorithm, we have $\alpha^*(t - 1) = \check{\eta}^*(\mathcal{V}^*(t - 1); \mathbb{A}^{t-1}) \leq \check{\eta}^*(\mathcal{V}^*)$. As shown in Lemma 3.2, the approximated function provides the global lower bound of the original optimization function, and the results are inside the feasible set of the approximate optimization function for the next iteration. Thus, $\check{\eta}^*(\mathcal{V}^*(t - 1); \mathbb{A}^{t-1}) \leq \check{\eta}(\mathcal{V}^*(t - 1); \mathbb{A}^t) \leq \check{\eta}^*(\mathcal{V}^*(t); \mathbb{A}^t)$. Therefore, $\alpha^*(t - 1) \leq \alpha^*(t)$. Moreover, due to the monotonically decreasing nature of $F(\alpha)$, $F^t(\alpha^*(t - 1)) \geq F^t(\alpha^*(t)) = 0$. $\square$

Given Lemma 3.3, the initial point in iteration t, i.e., $\alpha^0(t)$, in Algorithm 4 can be set at $\alpha^*(t - 1)$ rather than 0, so that the computing efficiency of the optimization algorithm can be further improved.

By now, the UAV computing energy efficiency maximization problem has been transformed into a solvable form. However, solving problem (3.28) is time-consuming due to multiple second order cone (SOC) constraints and requires the local information exchange between the UAV and devices. Therefore, we propose a distributed solution, in which devices maximize the amount of their offloaded computing tasks in parallel while the UAV aims to minimize its energy consumption. The original problem is decomposed into several sub-problems without losing optimality, and the UAV and devices solve the optimization problem cooperatively. Local information, such as the mobility of devices and the propulsion energy consumption function of the UAV, is not required to be shared among devices and the UAV.

We adopt the ADMM technique to decompose problem (3.28) [36]. The optimization solution is achieved in an iterative manner. Firstly, we introduce an auxiliary variable, $\mathbb{G}$, which is defined as:

$$\mathbb{G} = \begin{bmatrix} \ddot{\mathbf{Q}}_{1,1} & \cdots & \ddot{\mathbf{Q}}_{N,1} & W_{1,1} & \cdots & W_{N,1} \\ \vdots & \ddots & \vdots & \vdots & \ddots & \vdots \\ \ddot{\mathbf{Q}}_{1,K} & \cdots & \ddot{\mathbf{Q}}_{N,K} & W_{1,K} & \cdots & W_{N,K} \end{bmatrix}^T$$

where $\ddot{\mathbf{Q}}_{i,k}$ denotes the UAV location in time slot k expected by device i. Each device solves a part of the matrix $\mathbb{G}_i = [\ddot{\mathbf{Q}}_{i,1}, W_{i,1}; \ldots; \ddot{\mathbf{Q}}_{i,K}, W_{i,K}]$, and updates it to the UAV. Then, the UAV determines its trajectory, $\mathbf{Q}$, and overall computing load allocation according to the uploaded matrix $\mathbb{G}$. Denote the overall amount of computing load processed in slot k at the UAV by V_k, where $\mathbf{V} = [V_1; \ldots; V_K]$. The results determined by the UAV are summarized in matrix $\mathbb{H}$, where $\mathbb{H} = [\mathbb{I}_{(N \times 1)}\mathbf{Q}; \mathbf{V}]$, and $\mathbb{I}_{(N \times 1)}$ is a vector where all N entries are 1. By the end of the ADMM algorithm, the expected UAV trajectories should be unified and follow the flying constraints. The computing load should be allocated under the UAV computing capability. Thus, in the final optimal solution, the following constraint should be satisfied:

$$\mathbb{P}^T \mathbb{G} = \mathbb{H} \tag{3.29}$$

where

$$\mathbb{P} = \begin{bmatrix} \mathbb{I}_{(N \times N)} & \mathbf{0}_{(N \times 1)} \\ \mathbf{0}_{(N \times N)} & \chi \end{bmatrix}.$$

Vector χ represents the computing intensity for devices' tasks, where $\chi = [\chi_1; \ldots; \chi_N]$. The sub-matrices, $\mathbb{I}_{(N \times N)}$ and $\mathbf{0}_{(N \times N)}$, denote N-by-N identity matrix and zero matrix, respectively.

In addition, for the non-orthogonal channel model, we introduce another auxiliary variable, $e_{i,k}$, which denotes the summation of $\xi_{j,k}$ in all other devices except device i. This variable is used to decorrelate $\xi_{j,k}$ in (3.16) to facilitate an independent optimization process at each device. At the end of the optimization, $e_{i,k}$ should be equal to $\sum_{j \in \mathcal{I}/\{i\}} \xi_{j,k}$. For simplicity of presentation, we transform this constraint to

$$\frac{1}{N}(e_{i,k} + \xi_{i,k}) = \bar{\xi}_k \tag{3.30}$$

where $\bar{\xi}_k$ is the mean of $\{\xi_{1,k}, \ldots, \xi_{N,k}\}$. Then, the augmented Lagrangian function is formulated as follows:

$$\Gamma(\mathcal{V}_A) = -\sum_{k \in \mathcal{K}} \sum_{i \in \mathcal{I}} \check{R}_{i,k} + \alpha \Big[\sum_{k \in \mathcal{K}} \sum_{i \in \mathcal{I}} E_{i,k}^{C,U}(\mathbf{W}) + \sum_{k \in \mathcal{K}} \hat{E}_k^F \Big]$$
$$+ \mathrm{Tr}\{\mathbf{U}_1^T (\mathbb{P}^T \mathbb{G} - \mathbb{H})\} + \frac{\rho_1}{2} \|\mathbb{P}^T \mathbb{G} - \mathbb{H}\|_F^2$$

$$+\varpi \sum_{k\in\mathcal{K}}\sum_{i\in\mathcal{I}}\left\{U_{2,i,k}[\frac{1}{N}(e_{i,k}+\xi_{i,k})-\bar{\xi}_k]\right.$$

$$\left.+\frac{\rho_2}{2}[\frac{1}{N}(e_{i,k}+\xi_{i,k})-\bar{\xi}_k]^2\right\} \tag{3.31}$$

where $\|\cdot\|_F$ represents the Frobenius norm, $\mathcal{V}_A = \{\mathcal{V}, \mathbb{G}, \mathbb{H}, \mathbf{U}_1, \mathbf{U}_2\}$, $\mathbf{U}_1 \in \mathbf{R}^{(N+1)\times K}$ and $\mathbf{U}_2 \in \mathbf{R}^{N\times K}$ are Lagrange multipliers for the two auxiliary constraints, (3.29) and (3.30), respectively. Two parameters, ρ_1 and ρ_2, are penalty parameters. The parameter ϖ indicates the channel model, with $\varpi = 1$ for the non-orthogonal channel access scheme, and $\varpi = 0$ for the orthogonal channel access scheme.

Problem (3.28) can be separated into two sub-problems. The sub-problem solved in device i is organized as follows:

$$\min_{\mathcal{V}_1} \quad -\sum_{k\in\mathcal{K}}\check{R}_{i,k}+\mathrm{Tr}\{(\mathbf{U}_{1,i}^{n-1})^T\mathbb{P}_i^T\mathbb{G}_i\}+\frac{\rho_1}{2}\|\mathbb{P}_i^T\mathbb{G}_i-\mathbb{J}_i^{n-1}\|_F^2$$

$$+\varpi\left\{\frac{U_{2,i,k}^{n-1}(e_{i,k})}{N}+\frac{\rho_2}{2}(\frac{e_{i,k}-e_{i,k}^{n-1}}{N}+\theta_{i,k}^{n-1})^2\right.$$

$$\left.+\sum_{j\in\mathcal{I}/\{i\}}[-\frac{U_{2,j,k}^{n-1}\xi_{i,k}}{N}+\frac{\rho_2}{2}(\frac{\theta_{j,k}^{n-1}}{N-1}+\frac{\xi_{i,k}^{n-1}-\xi_{i,k}}{N})^2]\right\} \tag{3.32a}$$

s.t. $\quad \dfrac{(\|\ddot{\mathbf{Q}}_{i,k}-\mathbf{q}_{i,k}\|_2^2+H^2)n_0}{g_0} \le l_{i,k}, \forall i, k \tag{3.32b}$

$(3.7a),(3.7b),(3.10),(3.14e), (3.15h), (3.19),$

$(3.15b),$ if $\varpi = 0,$

(3.22), if $\varpi = 1.$

The sub-problem solved in the UAV is organized as follows:

$$\min_{\mathcal{V}_2} \quad \alpha[\sum_{k\in\mathcal{K}}\frac{\kappa V_k^3}{\Delta^2}+\sum_{k\in\mathcal{K}}\hat{E}_k^F]-\mathrm{Tr}\{(\mathbf{U}_1^n)^T\mathbb{H}\}+\frac{\rho_1}{2}\|\mathbb{P}^T\mathbb{G}^n-\mathbb{H}\|_F^2 \tag{3.33a}$$

s.t. $\quad \dfrac{V_k}{\Delta} \le f_{max}^U, \forall k, \tag{3.33b}$

$(3.14b),(3.14c),(3.15e), (3.21),(3.20).$

In (3.33), $(x)^{n-1}$ represents the variable x obtained in iteration $n-1$. The Lagrange multipliers, $\mathbf{U}_1$ and $\mathbf{U}_2$, are updated at each iteration as follows:

$$\mathbf{U}_1^n = \mathbf{U}_1^{n-1} + \rho_1(\mathbb{P}^T\mathbb{G}^n - \mathbb{H})^n \tag{3.34a}$$

$$\mathbf{U}_{2,i,k}^n = \mathbf{U}_{2,i,k}^{n-1} + \rho_2\theta_{i,k}^n \tag{3.34b}$$

where

$$\theta_{i,k}^n = \frac{1}{N}(e_{i,k}^n + \xi_{i,k}^n) - \bar{\xi}_k^n \tag{3.35}$$

represents the difference between the expected interference and the real interference. At iteration n, problem (3.32) is solved by each device individually. The optimization variable set $\mathcal{V}_1$ includes $\{\delta_{i,k}, W_{i,k}, \check{\mathbf{Q}}_{i,k}, \xi_{i,k}, \mathbf{1}, \check{\mathbf{R}}, e_{i,k}\}$ for all $k \in \mathcal{K}$. To decompose the auxiliary constraint (3.29) for each device i, we introduce sub-matrices $\mathbb{P}_i$, $\mathbb{H}_i$, and $\mathbf{U}_{1,i}$: $\mathbb{P}_i$ is a sub-matrix sliced from $\mathbb{P}$, where $\mathbb{P}_i = \mathbf{diag}\{1, \chi_i\}$; matrix $\mathbb{J}_i$ is obtained from the UAV information, where $\mathbb{J}_i^n = [\mathbf{Q}^n; \mathbf{V}^n/N + \chi_i W_i^n - \sum_{j \in \mathcal{I}} \chi_j \mathbf{W}_j^n/N]$; sub-matrix $\mathbf{U}_{1,i}$ is sliced from the dual variable, where $\mathbf{U}_{1,i} = [\mathbf{U}_1(i,:); \mathbf{U}_1(N+1,:)]$. Subsequently, problem (3.33) is solved by the UAV. The optimization variable set $\mathcal{V}_2$ includes $\{\mathbf{Q}, \omega, \mathbf{A}, \hat{\mathbf{E}}^F\}$.

Lemma 3.4 *If the initial value of $\{e^0, \xi^0, \mathbf{U}_1^0, \mathbf{U}_2^0\}$ is shared and unified among all devices and the UAV, only information from the UAV required for computing the sub-problem on the device side at each iteration is $\{\mathbb{J}_i^{n-1}, \theta^{n-1}\}$.*

Proof If the initial value is unified among the UAV and devices, the dual variables are not required to be shared and can be computed locally by the UAV and devices. For computing dual variable $\mathbf{U}_{1,i}$ at n, the following knowledge is required: updated global value $\mathbb{J}_i^{n-1}$, the historical value for local information $\mathbb{G}_i^{n-1}$, and the historical value of dual variable $\mathbf{U}_{1,i}^{n-1}$. Therefore, if $\mathbf{U}_{1,i}^0$ is identical to all devices and the UAV, $\mathbf{U}_{1,i}^n$ can be synchronized according to the historical value and the value from the global variable. Similarly, $\mathbf{U}_2$ can be evaluated by devices if the initial value is known. $\square$

Algorithm 5 ADMM Algorithm for solving problem (3.28)

1: Initialize variables $\{e^0, \xi^0, \theta^0, \mathbb{H}^0, \mathbb{G}^0\}$ and dual variables $\{\mathbf{U}_1^0, \mathbf{U}_2^0\}$. Loop index $n = 0$.
   Repeat
2: **For each device i:**
3: If $\varpi = 0$: Wait until receive updated $\mathbb{J}_i^{n-1}$.
4: If $\varpi = 1$: Wait until receive updated $\{\mathbb{J}_i^{n-1}, \theta^{n-1}\}$.
5: Calculate the dual variable $\mathbf{U}_{1,i}^{n-1} = \mathbf{U}_{1,i}^{n-2} + \rho_1(\mathbb{P}_i^T \mathbb{G}_i^{n-1} - \mathbb{J}_i^{n-1})$.
6: Calculate the dual variable $\mathbf{U}_2$ for all $i \in \mathcal{I}$ by (3.34b).
7: Solve problem (3.32).
8: If $\varpi = 0$: Send $\mathbb{G}_i^n$ to the server.
9: If $\varpi = 1$: Send $\{\mathbb{G}_i^n, e_i^n, \xi_i^n\}$ to the server.
10: **For the UAV-mounted server:**
11: Gather information from devices to form matrix $\mathbb{G}^n$.
12: Solve problem (3.33), and update $\mathbb{H}^n$.
13: Update dual variable $\mathbf{U}_1^n$ by (3.34a).
14: If $\varpi = 1$: Update variables $\theta_{i,k}^n \forall i, k$ by (3.35), and send the variables to devices.
15: $n = n + 1$.
   Until $|\Gamma^n(\mathcal{V}, \mathbb{G}, \mathbf{V}, \mathbf{U}_1, \mathbf{U}_2) - \Gamma^{n-1}(\mathcal{V}, \mathbb{G}, \mathbf{V}, \mathbf{U}_1, \mathbf{U}_2)| \leq \theta_3$.

Under the condition in Lemma 4, the distributed algorithm is given in Algorithm 5. In each optimization iteration, the devices compute and share matrix $\mathbb{G}$ with the UAV, and the UAV computes and shares matrix $\mathbb{J}$ to the devices. Meanwhile, when $\varpi = 1$, excepting contributing matrix $\mathbb{G}_i$, device i needs $e_{j,k}$ and $\xi_{j,k}$ from other devices $j \in \mathcal{I}/\{i\}$ to evaluate the interference.

By the problem decomposition, at the device side, each device aims only to maximize its own offloading data given the UAV trajectory computed by the UAV-mounted server and the interference environment in the previous iteration. At the UAV-mounted server side, the UAV aims to minimize energy consumption under the devices' expected UAV trajectories to collect enough workload. The trade-off between the received offloaded tasks and the energy consumption is controlled by parameter α which is updated in the ADMM algorithm loop. Meanwhile, the corresponding variables and constraints are split into two groups. This introduces three advantages. Firstly, local variables and parameters, such as device location and device offloading constraints, are not required to be uploaded to the UAV. Similarly, UAV's mechanical parameters and settings are not required to be shared with devices for offloading optimization. Secondly, less re-configuration is required when the UAV is replaced. Thirdly, the main computing load in solving the problem is from the SOC programming. The SOC constraints are now decomposed and solved by devices in parallel such that the computing efficiency can be improved. For the ADMM algorithm, in the orthogonal channel model, there are two main distributed blocks: the device side and the UAV side. The convergence of ADMM is guaranteed when the number of blocks is no more than two. In the non-orthogonal channel model, since each device is required to compute interference variable $e_{i,k}$ parallelly, convergence is not always guaranteed. Proximal Jacobian ADMM can be adopted to ensure the convergence, in which the proximal term, $\frac{\tau}{2}||x_i - x_i^k||^2$, is further combined in the primal problem of the current algorithm [37].

3.3.7 Proactive Trajectory Design Based on Spatial Distribution Estimation

So far, we have introduced the trajectory design and resource allocation for the scenario that all computing load information and device location are known. However, as some IoT devices are mobile [38], knowing their future positions during the upcoming computing cycle can be difficult. Moreover, devices needs to send the offloading requests at the beginning of the cycle. It means that the device may buffer the computing task until a new cycle begins, which introduces extra delay for waiting to send the request. Thus, the maximum queue delay may reach T seconds. To deal with the issues, in this subsection, we introduce an approach to estimate the spatial distribution of device locations in a cycle. The mobility of devices is predicted by an unsupervised learning tool, kernel density estimation method [39], and the computing load of each device is considered in a stochastic model correspondingly. The UAV trajectory is optimized via the estimated knowledge about ground devices. Thus, the UAV can collect the offloaded tasks of devices without requests in advance.

To estimate the location of devices, each device needs to report its current location periodically. The sampled location of device i is represented by q_i. We use the sampled location to estimate the spatial distribution of devices for the cycle, where the probability density function for the device at (x, y) is denoted as $f(x, y)$.

In order to compute $f(x, y)$, consider a small region, R, which is a rectangle area with side length of h_x and h_y, i.e., Parzen window. To count the number of devices within the region, we define the following function to indicate if device i is in the area:

$$C(q_i^x, q_i^y; R) = \begin{cases} 1, \text{ if } \max\{\frac{||q_i^x - x||}{h_x}, \frac{||q_i^y - y||}{h_y}\} \leq \frac{1}{2} \\ 0, \text{ otherwise} \end{cases} \tag{3.36}$$

where (x, y) is the central point of the area. Thus, for a large N value, the general expression for non-parametric density estimation is [39]

$$f(x, y) = \frac{1}{N h_x h_y} \sum_{i \in \mathcal{I}} C(q_i^x, q_i^y; R). \tag{3.37}$$

To establish a continuous estimation function, a smooth Gaussian kernel is applied,

$$\hat{f}(x, y) = \frac{1}{N \sqrt{h_x h_y}} \sum_{i \in \mathcal{I}} \frac{1}{2\pi} e^{-[\frac{(q_i^x - x)^2}{2h_x} + \frac{(q_i^y - y)^2}{2h_y}]} \tag{3.38}$$

which $\hat{f}(x, y)$ is the distribution obtained by Gaussian kernel estimation. In (3.38), h_x and h_y represent the bandwidth of the Gaussian kernel rather than the width of the Parzen window. To improve the estimation quality, the proper bandwidth, h_x and h_y, needs to be selected to minimize the error between the estimated density and the true density. Here, the maximum likelihood cross-validation method [39] is used to determine bandwidth h_x and h_y. The optimal bandwidth is

$$[h_x^*, h_y^*] = \arg\max \left\{ \frac{1}{N} \sum_{i \in \mathcal{I}} \log \hat{f}_{-i}(q_i^x, q_i^y) \right\} \tag{3.39}$$

where $\hat{f}_{-i}(q_i^x, q_i^y)$ is the estimated distribution in which device i is left out of the estimation.

In order to apply the estimated distribution into our proposed approach, we divide the working area of the UAV, $\mathcal{A}$, into $G \times G$ sub-areas. For each sub-area $\mathcal{A}_i$, there is a virtual device located at the center of the area. The virtual device carries all the computing tasks in the sub-area. It is assumed that the distribution of the task input data size and device spatial location are independent. The expected length of input bits for the tasks generated by a device is denoted by $\mathbb{E}[X]$. Thus, the expected length of computing bits generated inside the sub-area $\mathcal{A}_i$ is

$$\mathbb{E}[I_i] = \mathbb{E}[X]\mathbb{E}[N_i] = \mathbb{E}[X] \int_{(x,y)\in\mathcal{A}_i} \hat{f}(x,y)dxdy \qquad (3.40)$$

where $\mathbb{E}[N_i]$ denotes the expected number of devices in the sub-area $\mathcal{A}_i$. Our proposed approach can now solve the problem: In the new problem, there are G^2 virtual devices participating in the computing task offloading, and virtual device i has $\mathbb{E}[I_i]$ computing load to be completed in a cycle. The location of virtual device i is fixed at the center of the sub-area. For the orthogonal channel model, virtual device i shares a portion of $\mathbb{E}[N_i]/N$ of the channel bandwidth. As G increases, the performance of the estimation will be improved correspondingly.

3.4 Numerical Results

In this section, we evaluate the performance of our proposed optimization approach. The parameter setting is given in Table 3.2. The channel gain, g_0, is -70 dB. Let p represent the percentage of computing tasks offloaded to the server, i.e., $p = (\check{I}_i/I_i) * 100\%$. Devices have homogeneous offloading requirements in the simulation, i.e., E_i^T and p are identical for all devices. Consider both the non-orthogonal channel access scheme, denoted by "NO", and the orthogonal channel access scheme, denoted by "O". We consider the circular trajectory scheme as a benchmark, where the UAV moves on a circle within circular area centered at $(0.5, 0.5)$ km, and the radius is predefined. Two network scenarios are considered: a three-device scenario and a four-device scenario. In the three-device scenario, there are three devices located at $(0,1)$ km, $(1,1)$ km, and $(1,0)$ km, as shown in Fig. 3.2a. At the beginning of the cycle, the UAV moves from the location $(0,0)$ at an initial speed $(-10,0)$ m/s. By the end of the cycle, the UAV returns to the final designated position at $(0.5,0)$ km. In the four-device scenario, there are four devices located at randomly generated locations. The devices travel at constant speeds randomly

Table 3.2 Parameter setting for the computing scenario with three IoT devices

Parameter	Value	Parameter	Value
B	3 MHz	κ	10^{-28}
σ^2	-80 dBm/Hz	γ_1	0.0037
χ_i	1550.7	γ_2	500.206
Δ	1.5 s	H	100 m
a_{max}	50 m/s^2	P	100 mW
v_{max}	35 m/s	K	50

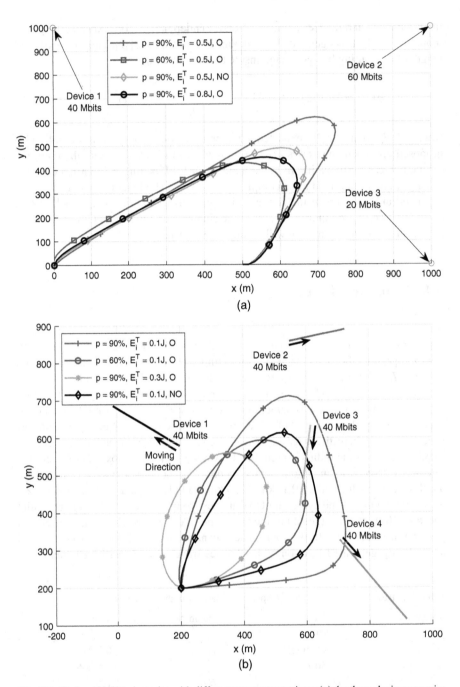

Fig. 3.2 Optimal UAV trajectories with different parameter settings: (**a**) the three-device scenario; (**b**) the four-device scenario with device mobility, where the solid straight lines represent device trajectories, and the arrows represent device moving directions

selected from $[-3,-3]$ m/s to $[3,3]$ m/s, as shown in Fig. 3.2b. The UAV moves from location $(200,200)$ m at an initial speed $(-10,0)$ m/s and returns to the initial position at the end of the cycle.

The UAV trajectory results obtained by the proposed approach are shown in Fig. 3.2. In the three-device case shown in Fig. 3.2a, the UAV takes most of the time moving towards and stays around the location of device 2 due to high computing task loads of the device. With a higher minimum offloading requirement, p, the UAV moves closer to devices in order to collect more offloading tasks. Similarly, with a lower maximum communication energy requirement, E_i^T, the UAV also moves closer to devices to reduce the device's offloading communication energy consumption. Moreover, since the non-orthogonal access method has a higher channel capacity, under the same condition, the trajectory of the non-orthogonal case shrinks to preserve the mechanical energy consumption as compared to the orthogonal channel case. Similar results can be obtained in the four-device case, as shown in Fig. 3.2b.

The comparison of the energy efficiency with different settings are shown in Fig. 3.3. In Fig. 3.3a and b, the x-axis represents the iteration number of the SCA-based algorithm loop. For both scenarios, with less task offloading demands, the energy efficiency improves due to the expanded optimization feasible set. In contrast, with more task offloading demands, the energy efficiency decreases significantly due to high energy consumption for the UAV to move closer to the devices.

For the three-device case, the ratio between the offloaded data amount and the overall computing data amount is shown in Fig. 3.4. The parameter setting for the indexes is given in Table 3.3, where the results using the proposed approach are shown in 1-6, and the results with the circular trajectory are shown in 7-9. For all scenarios, the proposed approach can meet the minimum offloading requirement, while the circular trajectory scheme cannot. Moreover, when the maximum communication energy requirement, E_i^T, increases, the UAV can collect more data even though its trajectory is further away from devices than in the case with a lower E_i^T value. The UAV collects extra offloaded tasks, which is beyond the devices' requirement, to improve its energy efficiency.

The trade-off between the maximum offloading energy, E_i^T, and the energy efficiency in the three-device case is shown in Fig. 3.5a. As E_i^T increases, the energy efficiency of the UAV increases at first and hits the ceiling with a high E_i^T. At that point, E_i^T is not the factor that limits the energy efficiency performance since all device's computing data is collected, as shown in Fig. 3.5c. When the energy efficiency reaches the maximum value, the UAV will find a path that has minimum energy consumption, given that all tasks are offloaded. Furthermore, our proposed approach can improve the energy efficiency significantly compared to the circular trajectory.

The magnitudes of the UAV acceleration and velocity in the three-device case are shown in Fig. 3.6a and b, respectively. The final velocity is constrained to be equal to the initial velocity. Note that the optimal velocity cannot be zero due to the

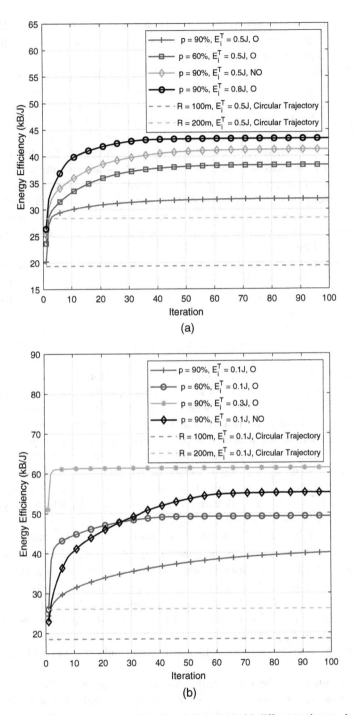

Fig. 3.3 Energy efficiency versus main loop iteration number with different trajectory designs: (**a**) the three-device scenario; (**b**) the four-device scenario with device mobility

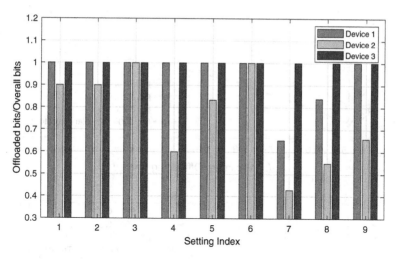

Fig. 3.4 The ratio between the offloaded task data amount and the overall computing task data amount generated by devices with different parameter settings

Table 3.3 Parameter setting for the computing scenarios in Fig. 3.4

Index	p	E_i^T	Index	p	E_i^T	Index	Radius	E_i^T
1	90%	0.5 J	4	60%	0.5 J	7	200 m	0.5 J
2	90%	0.8 J	5	60%	0.8 J	8	200 m	0.8 J
3	90%	1.1 J	6	60%	1.1 J	9	200 m	1.1 J

characteristic of fixed-wing UAV. With the lower maximum energy requirement, both magnitudes of acceleration and velocity increase, such that the UAV can move closer to devices. With the higher energy requirement, the fluctuation on velocity and acceleration decreases to reduce the propulsion energy consumption of the UAV.

The ratio of the actual allocated transmit power to the maximum power, $\delta_{i,k}$, for the three devices in a cycle is shown in Fig. 3.7a. Note that the overall offloading communication energy is limited. For the device with a high offloading demand, i.e., device 2, the ratio is maximized when the UAV moves close to it, while the ratio is minimized when the UAV moves away from it. The device tends to preserve the communication energy and starts the offloading only when the data rate is high. However, for device 3, the transmit power is allocated when the UAV is far away from the location of the device for two reasons: Firstly, the maximum communication energy of the device allows uploading the data even though the device transmission efficiency is low. Secondly, the UAV-mounted server prefers collecting the data in advance such that it can balance the computing load to reduce the computing energy cost.

The computing load allocation of the server in the three-device case is shown in Fig. 3.7b. Since the energy consumption increases cubically as the computing

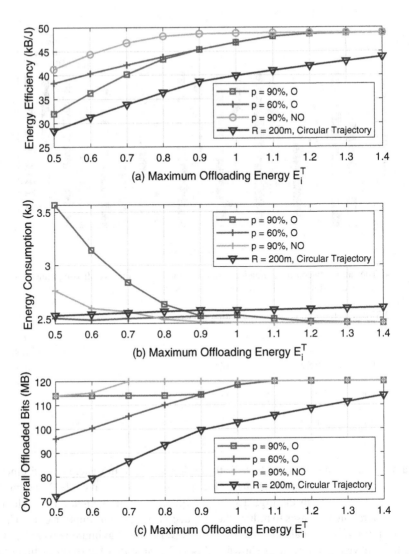

Fig. 3.5 (**a**) Energy efficiency versus the maximum offloading communication energy with different settings. (**b**) Overall energy consumption in a cycle versus the maximum offloading communication energy. (**c**) Overall offloaded bits in a cycle versus the maximum offloading communication energy

load in a unit time increases (based on (3.8) and (3.13)), the most energy-efficient computing load allocation policy is to balance the computing loads among time slots. As shown in Fig. 3.7b, the computing loads executed in a time slot may keep increasing at the beginning of a cycle for collecting the computing tasks and, then, remain at a fixed value for minimizing the computing energy consumption.

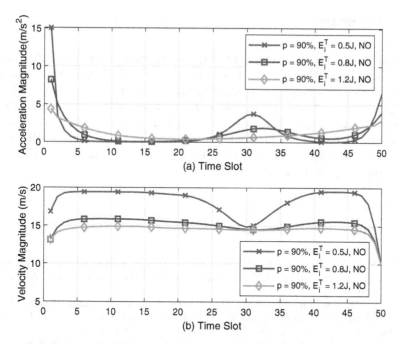

Fig. 3.6 (a) The acceleration of the UAV in the cycle. (b) The speed of the UAV in the cycle

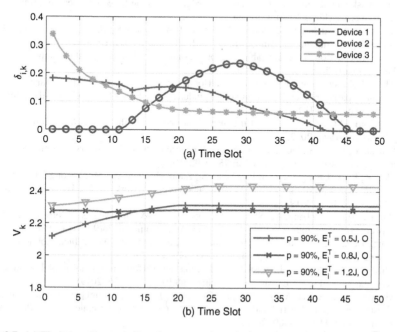

Fig. 3.7 (a) The transmit power allocation among three devices, where $p = 90\%$, and $E_i^T = 0.5$ J under orthogonal channel scenario. (b) The workload allocation with different settings

3.5   Summary

In this chapter, an optimization approach is presented to maximize the energy efficiency of a UAV-assisted MEC system, where the UAV trajectory design and resource allocation have been jointly considered. The non-convex and non-linear energy efficiency maximization problem has been solved in a distributed manner. Moreover, the device mobility estimation has been adopted to design a proactive UAV trajectory when the knowledge of device trajectory is limited. This study provides some insights on UAV optimal trajectory design for providing on-demand edge computing service for remote IoT devices.

References

1. Wu, Y., Shi, B., Qian, L.P., Hou, F., Cai, J., Shen, X.: Energy-efficient multi-task multi-access computation offloading via NOMA transmission for IoTs. IEEE Trans. Ind. Inf. **16**(7), 4811–4822 (2020)
2. Fu, S., Zhao, L., Ling, X., Zhang, H.: Maximizing the system energy efficiency in the blockchain based Internet of Things. In: 2019 IEEE Int. Conf. Commun. (ICC), pp. 1–6 (2019)
3. Hu, H., Wang, Q., Hu, R.Q., Zhu, H.: Mobility-aware offloading and resource allocation in an MEC-enabled IoT network with energy harvesting. IEEE Internet Things J. (2021, to appear)
4. Kong, X., Wang, K., Wang, S., Wang, X., Jiang, X., Guo, Y., Shen, G., Chen, X., Ni, Q.: Real-time mask identification for COVID-19: an edge computing-based deep learning framework. IEEE Internet Things J. (2021, to appear)
5. Mohamed, N., Al-Jaroodi, J., Jawhar, I., Noura, H., Mahmoud, S.: UAVFog: A UAV-based fog computing for Internet of Things. In: Proc. IEEE SmartWorld, Ubiquitous Intell. Comput., Adv. Trusted Computed, Scalable Comput. Commun., Cloud Big Data Comput., Internet People Smart City Innov, pp. 1–8 (2017)
6. Khan, W.Z., Aalsalem, M.Y., Khan, M.K., Hossain, M.S., Atiquzzaman, M.: A reliable Internet of Things based architecture for oil and gas industry. In: Proc. 19th Int. Conf. Adv. Commun. Technol., pp. 705–710 (2017)
7. Domingo, M.C.: An overview of the internet of underwater things. J. Netw. Comput. Appl. **35**(6), 1879–1890 (2012)
8. Samad, T., Bay, J.S., Godbole, D.: Network-centric systems for military operations in urban terrain: the role of UAVs. Proc. IEEE **95**(1), 92–107 (2007)
9. Shi, W., Li, J., Cheng, N., Lyu, F., Zhang, S., Zhou, H., Shen, X.: Multi-drone 3-D trajectory planning and scheduling in drone-assisted radio access networks. IEEE Trans. Veh. Technol. **68**(8), 8145–8158 (2019)
10. Wu, H., Tao, X., Zhang, N., Shen, X.: Cooperative UAV cluster-assisted terrestrial cellular networks for ubiquitous coverage. IEEE J. Sel. Areas Commun. **36**(9), 2045–2058 (2018)
11. Zeng, Y., Zhang, R., Lim, T.J.: Throughput maximization for UAV-enabled mobile relaying systems. IEEE Trans. Commun. **64**(12), 4983–4996 (2016)
12. Khabbaz, M.J., Antoun, J., Assi, C.: Modelling and performance analysis of UAV-assisted vehicular networks. IEEE Trans. Veh. Technol. **68**(9), 8384–8396 (2019)
13. Merwaday, A., Tuncer, A., Kumbhar, A., Guvenc, I.: Improved throughput coverage in natural disasters: Unmanned aerial base stations for public-safety communications. IEEE Veh. Technol. Mag. **11**(4), 53–60 (2016)
14. Hu, J., Zhang, H., Song, L.: Reinforcement learning for decentralized trajectory design in cellular UAV networks with sense-and-send protocol. IEEE Internet Things J. **6**(4), 6177–6189 (2019)

15. Enhanced LTE support for aerial vehicles. Tech. Rep. TR 36.777, Version 0.0.1, Release 15, 3GPP (2017)
16. Fu, S., Zhao, L., Su, Z., Jian, X.: UAV based relay for wireless sensor networks in 5G systems. Sensors **18** (2018)
17. Zhou, Y., Cheng, N., Lu, N., Shen, X.: Multi-UAV-aided networks: aerial-ground cooperative vehicular networking architecture. IEEE Veh. Technol. Mag. **10**(4), 36–44 (2015)
18. Cheng, N., Xu, W., Shi, W., Zhou, Y., Lu, N., Zhou, H., Shen, X.: Air-ground integrated mobile edge networks: architecture, challenges, and opportunities. IEEE Commun. Mag. **56**(8), 26–32 (2018)
19. Huo, Y., Dong, X., Lu, T., Xu, W., Yuen, M.: Distributed and multi-layer UAV networks for next-generation wireless communication and power transfer: a feasibility study. IEEE Internet Things J. **6**(4), 7103–7115 (2019)
20. Mao, Y., You, C., Zhang, J., Huang, K., Letaief, K.B.: A survey on mobile edge computing: the communication perspective. IEEE Commun. Surv. Tut. **19**(4), 2322–2358 (2017)
21. Hu, Q., Cai, Y., Yu, G., Qin, Z., Zhao, M., Li, G.Y.: Joint offloading and trajectory design for UAV-enabled mobile edge computing systems. IEEE Internet Things J. **6**(2), 1879–1892 (2019)
22. Wu, Q., Zeng, Y., Zhang, R.: Joint trajectory and communication design for multi-UAV enabled wireless networks. IEEE Trans. Wireless Commun. **17**(3), 2109–2121 (2018)
23. Zeng, Y., Zhang, R.: Energy-efficient UAV communication with trajectory optimization. IEEE Trans. Wireless Commun. **16**(6), 3747–3760 (2017)
24. Tang, F., Fadlullah, Z.M., Kato, N., Ono, F., Miura, R.: AC-POCA: Anticoordination game based partially overlapping channels assignment in combined UAV and D2D-based networks. IEEE Trans. Veh. Technol. **67**(2), 1672–1683 (2018)
25. Garg, S., Singh, A., Batra, S., Kumar, N., Yang, L.T.: UAV-empowered edge computing environment for cyber-threat detection in smart vehicles. IEEE Net. **32**(3), 42–51 (2018)
26. Messous, M., Sedjelmaci, H., Houari, N., Senouci, S.: Computation offloading game for an UAV network in mobile edge computing. In: 2017 IEEE Int. Conf. Commun. (ICC), pp. 1–6 (2017)
27. Jeong, S., Simeone, O., Kang, J.: Mobile edge computing via a UAV-mounted cloudlet: Optimization of bit allocation and path planning. IEEE Trans. Veh. Technol. **67**(3), 2049–2063 (2018)
28. Tang, F., Fadlullah, Z.M., Mao, B., Kato, N., Ono, F., Miura, R.: On a novel adaptive UAV-mounted cloudlet-aided recommendation system for LBSNs. IEEE Trans. Emerg. Topics Comput. **7**(4), 565–577 (2019)
29. Cheng, N., Lyu, F., Quan, W., Zhou, C., He, H., Shi, W., Shen, X.: Space/aerial-assisted computing offloading for IoT applications: a learning-based approach. IEEE J. Sel. Areas Commun. **37**(5), 1117–1129 (2019)
30. Wang, Y., Sheng, M., Wang, X., Wang, L., Li, J.: Mobile-edge computing: partial computation offloading using dynamic voltage scaling. IEEE Trans. Commun. **64**(10), 4268–4282 (2016)
31. Wang, F., Xu, J., Wang, X., Cui, S.: Joint offloading and computing optimization in wireless powered mobile-edge computing systems. IEEE Trans. Wireless Commun. **17**(3), 1784–1797 (2018)
32. Yuan, W., Nahrstedt, K.: Energy-efficient soft real-time CPU scheduling for mobile multimedia systems. SIGOPS Oper. Syst. Rev. **37**(5), 149–163 (2003)
33. Li, H., Ota, K., Dong, M.: Learning IoT in edge: Deep learning for the Internet of Things with edge computing. IEEE Netw. **32**(1), 96–101 (2018)
34. Lipp, T., Boyd, S.: Variations and extension of the convex–concave procedure. Optim. Eng. **17**(2), 263–287 (2016)
35. Dinkelbach, W.: On nonlinear fractional programming. Manage. Sci. **13**(7), 492–498 (1967)
36. Boyd, S., Parikh, N., Chu, E., Peleato, B., Eckstein, J.: Distributed optimization and statistical learning via the alternating direction method of multipliers. Found. Trends Mach. Learn. **3**(1), 1–122 (2011)
37. Deng, W., Lai, M., Peng, Z., Yin, W.: Parallel multi-block ADMM with O(1/k) convergence. J. Sci. Comput. **71**(2), 712–736 (2017)

38. Hakiri, A., Berthou, P., Gokhale, A., Abdellatif, S.: Publish/subscribe-enabled software defined networking for efficient and scalable IoT communications. IEEE Commun. Mag. **53**(9), 48–54 (2015)
39. Cao, R., Cuevas, A., Manteiga, W.G.: A comparative study of several smoothing methods in density estimation. Comput. Stat. Data Anal. **17**(2), 153–176 (1994)

Chapter 4
Collaborative Computing for Internet of Vehicles

In this chapter, a collaborative edge computing framework is presented to reduce computing service latency and improve service reliability for vehicular networks. First, a task partition and scheduling algorithm is proposed to decide the workload allocation and the execution order of tasks offloaded to edge servers. Second, an artificial intelligence based collaborative computing approach is developed to determine the task offloading, computing, and result delivery policy for vehicles. Specifically, the offloading and computing problem is formulated as a Markov decision process. A deep reinforcement learning technique, i.e., deep deterministic policy gradient, is adopted to find the optimal solution for a complex urban transportation network. Our approach minimizes the service cost, which includes computing service latency and service failure penalty, via the optimal workload assignment and server selection in collaborative computing. Simulation results show that the proposed learning-based collaborative computing approach can adapt to a highly dynamic environment and perform well.

4.1 Background on Internet of Vehicles

Vehicular communication networks have drawn significant attention from both academia and industry in the past decade. Conventional vehicular networks aim to improve the driving experience and safety via data exchange using vehicle-to-everything (V2X) communications. As the communication capacity keeps increasing in the 5G era, the concept of vehicular networks has been extended to Internet of Vehicles (IoV). Via the integration of communication and computing in vehicular networks, IoV targets interactive vehicular applications, such as assisted/autonomous driving, platooning, urban traffic management, and on-board infotainment services [1, 2].

© The Author(s), under exclusive license to Springer Nature Switzerland AG 2021
J. Gao et al., *Connectivity and Edge Computing in IoT: Customized Designs and AI-based Solutions*, Wireless Networks,
https://doi.org/10.1007/978-3-030-88743-8_4

The IoV is anticipated to address the challenges of modern transportation networks. Vehicles can play the role of intelligent sensors to sense the conditions of a transportation network, such as traffic jams and accidents. Within the IoV, vehicular networks can become a platform to gather and analyze sensor data and provide information for vehicles, such as the HD maps for autonomous driving. The information can be analyzed by vehicles to facilitate intelligent transportation, such as traffic congestion relief, fuel consumption, and pollution reduction. Moreover, on-board applications can enhance the traveling experience of passengers. Therefore, the IoV can lead to a revolution in various sectors of the automotive industry, including vehicle manufacturing, energy, automation, and software.

Although the IoV is promising, realizing IoV applications faces challenges. One of the obstacles is the limited on-board computing capability at vehicles. For example, a self-driving car with ten high-resolution cameras may generate 2 gigapixels of data per second, while 250 trillion computing operations per second are required to process the data promptly [3]. Supporting such compute-intensive applications on vehicular terminals is energy-inefficient and time-consuming. The MEC is a possible solution for supporting low-latency and energy-efficient computing services for vehicles [4–6]. Via vehicle-to-infrastructure (V2I) communications, resource-constrained vehicles can offload their compute-intensive tasks to highly capable edge servers co-located with roadside units (RSUs) for task processing. Meanwhile, compared with cloud computing, the delay caused by task offloading can be significantly reduced in MEC due to the proximity of edge servers to vehicles [7]. Consequently, applications that require high computing capabilities, such as path navigation, video stream analytics, and objective detection, can be enabled in IoV by MEC [8]. Despite the advantage brought by MEC, new challenges have emerged in task offloading and computing. In vehicular networks, the high mobility of users leads to intermittent communication links [9], while an edge server may take some time to execute the offloaded tasks. Due to the non-negligible computing time and the limited communication range of vehicles, a vehicle may travel out of the coverage area of an edge server during a computing session, resulting in a service disruption. Therefore, proper computing resource management is required to support seamless MEC in highly dynamic vehicular networks.

4.2 Connectivity Challenges for MEC

The main challenge in supporting IoV applications is to maintain the connectivity between an edge server and vehicles during a computing session. In this section, we review existing research on computing offloading and introduce two main approaches for seamless edge computing in IoV.

4.2.1 Server Selection for Computing Offloading

Computing offloading, i.e., a vehicle offloading computing tasks to an edge server, is the first step of edge computing. When the communication coverage of edge servers overlaps, vehicles may be able to connect to multiple edge servers. An edge server with larger communication coverage can potentially maintain a longer connection duration for a vehicle, which can reduce the chance of service disruption and the communication overhead for migrating computing services across edge servers. Therefore, the mobility of vehicles impacts computing offloading policy, while associating multiple vehicles with differentiated mobility and locations to multiple edge servers is a challenging problem.

The problem of computing offloading has been investigated in many research works in the context of vehicular networks [10–12], where the main objective is to minimize service delay by edge server selection. In [13–15], machine learning techniques are used to make offloading decisions for vehicles via predicting their trajectories. In [13], Sun et al. focus on task offloading and execution utilizing computing resources on vehicles, i.e. vehicular edge. An online learning algorithm, i.e., multi-armed bandit, is utilized to determine the association between vehicles and edge servers. In [14], Ning et al. apply a Deep Reinforcement Learning (DRL) approach to jointly allocate communication, caching, and computing resources in a dynamic vehicular network.

Even with proper edge server selection for computing offloading, it is possible that vehicles travel out of the communication coverage of an edge server during their computing sessions. To ensure that vehicles can obtain their computing results at the end of a computing session, multiple edge servers may need to collaborate to provide seamless connectivity for vehicles.

4.2.2 Service Migration

To support reliable computing services for vehicles with high mobility, service migration is a potential solution. Service migration aims to adjust the server-vehicle association when vehicles leave a server's communication range. According to the trajectory of a vehicle, an ongoing computing service can be moved to another edge server that will cover the vehicle in near future. Migration decisions are made according to a variety of factors, including the communication link quality, computing capability of edge servers, and vehicle mobility. In [16, 17], service migration schemes among federated cloud data centers are presented, which can be extended to the MEC system. An ongoing service can be migrated to another server as the corresponding vehicle moves out of range. Based on the random walk model, a Markov decision process (MDP) for determining the migration policy is developed in [18]. The MDP method is to make a proactive decision on whether a service should be migrated according to the mobility of vehicles. Similar proactive service migration strategies are investigated in [19] and [20].

Overall, computing service migration provides a practical solution for implementing MEC in a highly dynamic environment. However, it increases the complexity of resource management. Specifically, when a vehicle is beyond the communication range of an edge server, the edge server needs to determine whether to interrupt and migrate the ongoing service to another edge server or not. If yes, a proper edge server should be selected to receive the migrated service according to the vehicle trajectory. Otherwise, a proper transmission policy is required to maintain the connectivity between the vehicle and the edge server so that the computing result can be delivered at the end of the session. An effective migration policy should adapt to the dynamics in channel condition, server computing capability, vehicle location, migration overhead, etc. [21], while obtaining such a policy can be difficult due to high vehicle mobility. A potential approach is proactive service migration using Artificial Intelligence (AI) techniques based on learning the vehicles' mobility and channel conditions.

4.2.3 Cooperative Computing

Different from service migration, which enhances service reliability by migrating computing services according to the vehicle's trajectory, service cooperation improves MEC service reliability by accelerating task processing. A computing task can be divided and computed by multiple servers in parallel or offloaded to one server with high computing capability [22, 23]. With cooperative computing, a computing task can be forwarded to an edge server out of the vehicle's communication range. Compared to service migration, in which each edge server executes tasks offloaded only by vehicles in its communication coverage, service cooperation allows edge servers to process tasks offloaded by vehicles out of their coverage for reducing the overall computing delay.

Service cooperation has been studied in [4, 24] and [25]. The works [4] and [24] consider that vehicles divide and offload the computing tasks to multiple servers according to their trajectories. Vehicle-to-vehicle communication is used to disseminate computing results if edge servers cannot connect with vehicles at the end of a computing session. In [25], the authors utilize neural networks to predict computing demands in a vehicular network, and MEC servers are clustered to compute offloaded tasks cooperatively. Nevertheless, wireless transmission among edge servers or vehicles in the task offloading process can result in significant transmission delay and communication resource consumption. The tradeoff between communication overhead and computing capability increases the complexity of server assignment in collaborative computing.

4.3 Computing Task Partition and Scheduling for Edge Computing

We present a computing collaboration framework to provide reliable low-latency computing in an MEC-enabled vehicular network. Once an edge server receives the computing tasks offloaded by a vehicle, it may partially or fully distribute the computing workload to another edge server to reduce computing latency. Furthermore, by selecting proper edge servers to deliver the computing results, vehicles are able to obtain computing results without service disruption caused by mobility. Under this framework, we present a novel task offloading and computing approach that reduces the overall computing service latency and improves service reliability. To achieve this objective, we firstly formulate a task partition and scheduling optimization problem, which allows all received tasks in the network to be executed with minimized latency given the offloading strategy. A heuristic task partition and scheduling approach is developed to obtain a near-optimal solution of the non-convex integer problem. In addition, we formulate the radio and computing resource allocation problem as an MDP. An AI approach, DRL, is adopted to find a proactive offloading policy for vehicles by evaluating the MDP. Specifically, a convolutional neural network (CNN) based DRL is developed to handle the high-dimensional state space, and a deep deterministic policy gradient (DDPG) algorithm is adopted to handle the high-dimensional action space.

4.3.1 Collaborative Edge Computing Framework

An MEC-enabled vehicular network is illustrated in Fig. 4.1. A row of RSUs, equipped with computing resources, provide seamless communication and computing service coverage for vehicles on the road. An RSU can communicate with other RSUs within its communication range via wireless links. The set of RSUs is denoted by $\mathcal{R}$, where the index of RSUs is denoted by $r \in \mathcal{R}$. We divide each one-way road into several zones with equal length, where the set of zones is denoted by $\mathcal{Z}$. We divide time into time slots, where the index and set of time slots are denoted by t and $\mathcal{T}$, respectively. The index of the zones is denoted by $z = (a, b) \in \mathcal{Z}$. The parameters, a and b, represent the index of roads and the index of segments on the road, respectively, where $a \in \{1, \ldots, A\}$, and $b \in \{1, \ldots, B\}$. As the vehicle drives through the road, it traverses the zones consecutively. We assume that all vehicles in the same zone follow the same offloading and computing policy.[1] For simplicity, we aggregate tasks from vehicles in each zone and refer to the tasks offloaded by vehicles in zone z as task z. We suppose that the vehicle will not travel out of a zone

---

[1] The accuracy of vehicle locations improves as the length of the zone is reduced. In consideration of the length of a car, the length of a zone is larger than 5 m.

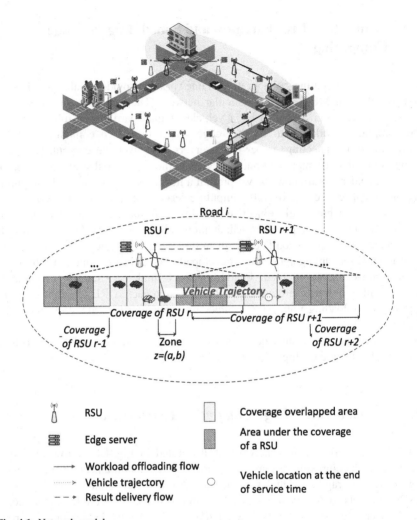

Fig. 4.1 Network model

during the time duration of a time slot, and vehicles can complete the offloading process of a task generated in a zone before it travels out of the zone. We assume that a global controller has full knowledge of the transportation network and makes offloading and computing decisions for all the vehicles in a centralized manner. In our model, a computing session for a task includes three steps:

(1) Offloading—When a computing task is generated at a vehicle, the vehicle selects an RSU, which is under its communication range, and offloads the computing data of the task to the RSU immediately. In the example shown in Fig. 4.1, RSU r is selected for computing tasks offloaded by the target vehicle. Such RSU is referred to as *receiver RSU* for the task;

(2) Computing—After the computing task is fully offloaded, receiver RSU either processes the whole computing task or selects another RSU to share the computing load. The RSU, which is selected to process the task collaboratively with receiver RSU, is referred to as *helper RSU* for the task;

(3) Delivering—A vehicle may travel out of the communication range of its receiver RSU. Therefore, the controller will select an RSU, which can connect with the vehicle at the end of computing session, to gather and transmit computing results. The RSU is referred to as *deliver RSU*. For efficiency, we limit deliver RSU to be either receiver RSU or helper RSU of the task. In Fig. 4.1, RSU $r + 1$ behaves as both helper RSU and deliver RSU for the computing task offloaded by the vehicle.

To reduce the decision space in task offloading and computing, instead of providing the offloading and computing policy to individual vehicles, we consider location-based offloading and computing policy. Denote the set of vehicles in zone z and time slot $t \in \mathcal{T}$ as $\mathcal{V}_{z,t}$. The offloading decision for vehicles in zone z and time slot t is represented by a vector $\boldsymbol{\alpha}_{z,t} \in \mathbb{Z}_+^{|\mathcal{R}|}$, where $\sum_{r=1}^{|\mathcal{R}|} \alpha_{z,r,t} = 1$. The element, $\alpha_{z,r,t}$, is 1 if RSU r is selected as receiver RSU for the vehicles in zone z and time slot t, and 0 otherwise. Similarly, the collaborative computing decision for vehicles in zone z and time slot t is represented by vector $\boldsymbol{\beta}_{z,t} \in \mathbb{Z}_+^{|\mathcal{R}|}$, where $\sum_{r=1}^{|\mathcal{R}|} \beta_{z,r,t} = 1$. The element, $\beta_{z,r,t}$, is 1 if RSU r is selected as helper RSU for the vehicles in zone z and time slot t, and 0 otherwise. In addition, the decision on result delivery is denoted by a binary variable, $\gamma_{z,r,t}$, where $\gamma_{z,r,t}$ is 1 if the computing results are delivered by RSU r for task z in time slot t and 0 otherwise. The system cost comes from the service delay and service failure.

4.3.2 Service Delay

We adopt the task partition technique during task processing [26, 27]. A task offloaded by vehicles in a zone can be divided and processed by receiver RSU and helper RSU cooperatively. At each time slot, receiver RSU, helper RSU, and deliver RSU are selected for each zone according to the task offloading policy. Vehicles in a zone offload the task to selected receiver RSU for the zone. Once a receiver RSU receives the offloaded tasks, it immediately divides the tasks and offloads a part of each task to helper RSU selected for the zone. We denote the computing delay of task z corresponding to receiver or helper RSU r in time slot t as $M_{z,r,t}$. As shown in Fig. 4.2, the computing delay includes task offloading delay, queuing delay, and processing delay. Since the amount of output data is usually much smaller compared to the amount of input data, we neglect the transmission delay in result delivery [12, 28]. We assume that the workload for partitioning tasks is much smaller than the workload for processing offloaded tasks, and thus we neglect task partition delay.

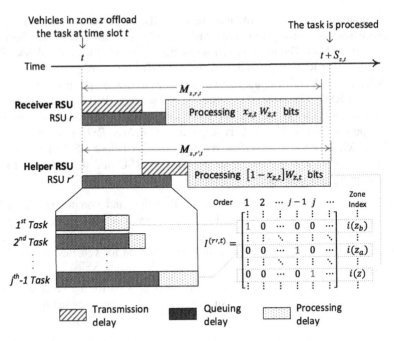

Fig. 4.2 An example of the task offloading and computing process

Firstly, task offloading comprises two steps: offloading tasks from vehicles to their receiver RSU and offloading the partial workload from receiver RSU to helper RSU. According to the propagation model in 3GPP standards [29], the path loss between a transmitter and a receiver with distance d (km) can be computed as:

$$L(d) = 40(1 - 4 \times 10^{-3} D^{hb}) \log_{10} d - 18 \log_{10} D^{hb} \qquad (4.1)$$
$$+ 21 \log_{10} f + 80 \text{ (dB)}$$

where f is the carrier frequency in MHz, and D^{hb} represents the antenna height in meter. We do not consider the shadowing effect of the channel. Denote the distance between the center point of zone z and the location of RSU r as $D_{z,r}$, and the distance between RSU r and r' as $D_{r,r'}$. The data rate for vehicles in zone z offloading task to RSU r is

$$r_{z,r} = B^Z \log_2 \left(1 + \frac{P^V 10^{-L(D_{z,r})/10}}{\sigma_v^2}\right) \qquad (4.2)$$

where σ_v^2 denotes the power of the received Gaussian noise in the V2I channel, P^V represents the vehicle transmit power, and B^Z represents the bandwidth reserved for vehicles in a zone. At the receiver RSU for task z, a signal-to-noise ratio threshold should be satisfied, given by

$$\frac{P^V 10^{-L(D_{z,r})/10}}{\sigma_v^2} \geq \alpha_{z,r,t}\delta^O, \forall t, z, r \tag{4.3}$$

where δ^O is the signal-to-noise ratio threshold for data offloading. Assume that vehicles in a zone are scheduled to offload the tasks successively, and the channel is time-invariant in the duration of any computing task offloading. The transmission delay for offloading the computing data in zone z to receiver RSU is

$$A_{z,t} = \sum_{r \in \mathcal{R}} \frac{\alpha_{z,r,t} W_{z,t}}{r_{z,r}} \tag{4.4}$$

where $W_{z,t}$ represents the overall computing data generated by vehicles in zone z, i.e., task z, and time slot t. In addition, the data rate between RSU r and RSU r' for forwarding the computing data offloaded from a zone is

$$r_{r,r'} = B^R \log_2 \left(1 + \frac{P^R 10^{-L(D_{r,r'})/10}}{\sigma_r^2}\right) \tag{4.5}$$

where σ_r^2 represents the power of received Gaussian noise power in the RSU to RSU channel, P^R represents the RSU transmit power, and B^R represents the bandwidth reserved for forwarding data offloaded from a zone. In data forwarding, the signal-to-noise constraint is required to be satisfied,

$$\frac{P^R 10^{-L(D_{r,r'})/10}}{\sigma_r^2} \geq \beta_{z,r',t}\delta^O, \forall t, z, r, r'. \tag{4.6}$$

For computing task z in time slot t, the portion of workload to be processed by receiver RSU and helper RSU is denoted by $x_{z,t}$ and $1 - x_{z,t}$, respectively. Thus, the delay for forwarding the data to deliver RSU is

$$F_{z,t} = \sum_{r \in \mathcal{R}} \sum_{r' \in \mathcal{R}} \frac{\alpha_{z,r,t}\beta_{z,r',t}(1 - x_{z,t})W_{z,t}}{r_{r,r'}}. \tag{4.7}$$

Furthermore, after the task is offloaded to edge servers, a queuing delay may be experienced. Let set $\mathcal{Z}^{r,t}$ denote the zones which have tasks offloaded to RSU r, i.e., $\{z|\alpha_{z,r,t} = 1\} \cup \{z|\beta_{z,r,t} = 1\}$, and let $i(z)$ represent the index of zone z in set $\mathcal{Z}^{r,t}$. We denote $N_{r,t}$ as the number of tasks offloaded in time slot t and assigned to RSU r, where $N_{r,t} = \sum_z \alpha_{z,r,t} + \beta_{z,r,t}$. Then, a matrix, $\mathbb{I}^{(r,t)} \in \mathbb{Z}_+^{N_{r,t} \times N_{r,t}}$, is defined to denote the processing order of tasks offloaded to RSU r in time slot t, where $I_{i(z),j}^{(r,t)} = 1$ if the task offloaded from zone z is scheduled as the j-th task to be processed among all the tasks offloaded in the same time slot. As shown in Fig. 4.2, the queuing delay of a task depends on the computing delay of the task schedule priorly. For the first task to be processed among the tasks offloaded in time slot t, the queuing delay stems from the computing delay for the tasks offloaded in previous time slots. Thus, the queuing delay of task z in RSU r is given by

$$
U_{z,r,t} = \begin{cases} \hat{U}_{r,t}, & \text{if } I^{(r,t)}_{i(z),1} = 1, \\ \sum_{z'} \sum_j I^{(r,t)}_{i(z),j} I^{(r,t)}_{i(z'),j-1} M_{z',r,t}, & \text{otherwise.} \end{cases} \tag{4.8}
$$

In (4.8), $\hat{U}_{r,t}$ represents the latency for finishing the tasks offloaded in previous time slots $\{1, \ldots, t-1\}$, where

$$
\hat{U}_{r,t} = \max \left\{ \sum_{z'} I^{(r,t)}_{i(z'),N_{r,t-1}} M_{z',r,t-1} - \epsilon, 0 \right\} \tag{4.9}
$$

and ϵ is the length of a time slot.

We consider that data transmission and task processing are in parallel. After the task is offloaded and other tasks scheduled priorly are completed, the task can be processed by the dedicated server. The delay for processing task z offloaded to RSU r in time slot t is given by

$$
G_{z,r,t} = \frac{\chi W_{z,t} [\alpha_{z,r,t} x_{z,t} + \beta_{z,r,t} (1 - x_{z,t})]}{C_r} \tag{4.10}
$$

where C_r denotes the computing capability (CPU-cycle frequency) of RSU r, and χ denotes the number of computing cycles needed to process 1 bit of data.

Given the offloading delay, queuing delay, and processing delay, the computing delay for task z on RSU r is

$$
M_{z,r,t} = \max \left\{ A_{z,t} + \beta_{z,r,t} F_{z,t}, U_{z,r,t} \right\} + G_{z,r,t}. \tag{4.11}
$$

Denote the overall service delay for the task offloaded from zone z in time slot t by $S_{z,t}$. As shown in Fig. 4.2, the overall service delay depends on the longest computing delay between receiver RSU and helper RSU. Thus, we have

$$
S_{z,t} = \max \left\{ \sum_r \alpha_{z,r,t} M_{z,r,t}, \sum_r \beta_{z,r,t} M_{z,r,t} \right\}. \tag{4.12}
$$

4.3.3 Service Failure Penalty

The mobility of vehicles brings uncertainty in result downloading. Service failure may occur if a vehicle is out of the coverage of its deliver RSU during the computing session. Denote the zone that vehicle v is located when its computing result is delivered by m_v, i.e., the location of vehicle $v \in V_{z,t}$ in time slot $t + S_{z,t}$. Also, we denote the signal-to-noise ratio threshold for result delivering as δ^D. Let $\mathbf{1}_{z,t}$ indicate whether the computing service for task z offloaded in time slot t is successful or not, where

$$\mathbf{1}_{z,t} = \begin{cases} 1, & \text{if } P^R 10^{-L(D_{m_v,r})/10} \geq \sigma_r^2 \gamma_{z,r,t} \delta^D, \forall v \in \mathcal{V}_{z,t} \\ 0, & \text{otherwise.} \end{cases} \tag{4.13}$$

4.3.4 Problem Formulation

The objective is to minimize the weighted sum of the overall computing service delay for vehicles and service failure penalty. The objective function is as follows:

$$\min_{\substack{\{\alpha,\beta,\gamma,\mathbf{x}, \\ \{\mathbf{I}^{(r,t)},\forall r,t\}\}}} \lim_{T\to\infty} \frac{1}{T} \sum_{t=0}^{T-1} \sum_{z\in\mathcal{Z}} \left\{ S_{z,t}\mathbf{1}_{z,t} + \lambda W_{z,t}(1 - \mathbf{1}_{z,t}) \right\} \tag{4.14a}$$

$$\text{s.t. } (4.3), (4.6), \tag{4.14b}$$

$$\sum_{r\in\mathcal{R}} \alpha_{z,r,t} = 1, \sum_{r\in\mathcal{R}} \beta_{z,r,t} = 1, \sum_{r\in\mathcal{R}} \gamma_{z,r,t} = 1 \tag{4.14c}$$

$$\sum_{i=1}^{N_{r,t}} I_{i,j}^{(r,t)} = 1, \sum_{j=1}^{N_{r,t}} I_{i,j}^{(r,t)} = 1 \tag{4.14d}$$

$$0 \leq x_{z,t} \leq 1, \tag{4.14e}$$

$$\alpha_{z,t}, \beta_{z,t} \in \mathbb{Z}_+^{|\mathcal{R}|}, \tag{4.14f}$$

$$\mathbf{I}^{(r,t)} \in \mathbb{Z}_+^{N_{r,t} \times N_{r,t}} \tag{4.14g}$$

where λ represents per-unit penalty for the case when a computing service fails.

There are two types of optimization variables: edge server selection for task offloading and computing, i.e., $\{\alpha, \beta, \gamma\}$, and the computing policy for the offloaded tasks, including task partition, i.e., $\mathbf{x}$, and task execution order, i.e., $\{\mathbf{I}^{(r,t)}, \forall r, t\}$.

It can be seen that Problem (4.14) is a mixed-integer nonlinear optimization problem. Solving the problem by conventional optimization methods is challenging. Furthermore, the number of variables in the problem is too large to apply model-free techniques directly. Taking the variables of task execution order as an example, i.e., $\mathbf{I}^{(r,t)}$, there are $N_{r,t} \times N_{r,t}$ decisions to be determined for a server in a time slot. The number of combinations of task execution order is at least $(|\mathcal{Z}|/|\mathcal{R}|)! \times |\mathcal{R}| \times |\mathcal{T}|$, in which tasks are evenly assigned to servers and each task is processed by only one server. Thus, to solve the problem in a scalable manner, we divide Problem (4.14) into two sub-problems: (i) task partition and scheduling problem, and (ii) edge server selection problem. Specifically, in the task partition and scheduling problem, we aim to obtain the optimal task partition ratio and the execution order to minimize the computing latency given an edge server selection decision, i.e., $\{\alpha, \beta\}$. After that, we re-formulate the edge server selection problem as an MDP and utilize the DRL technique to obtain the optimal edge server selection decision.

4.3.5 Task Partition and Scheduling

In each time slot, an edge server receives tasks offloaded by vehicles in different zones. The overall computing service delay for a task depends on the task execution order in edge servers. In addition, the workload of a task can be divided and offloaded to two edge servers, i.e., receiver and helper RSUs. The workload allocation of a task also affects the overall computing service delay. Therefore, we study task partition and scheduling to minimize the service delay given a edge server selection policy $\{\alpha, \beta\}$. Based on Problem (4.14), the delay minimization problem is formulated as follows:

$$\min_{\mathbf{x}, \{\mathbf{I}^{(r,t)}, \forall r, t\}} \sum_{z \in \mathcal{Z}} S_{z,t} \tag{4.15a}$$

$$\text{s.t.} \quad (4.14d), (4.14e), (4.14g). \tag{4.15b}$$

Problem (4.15) is a mixed-integer programming, which involves a continuous variable, $\mathbf{x}$, and an integer matrix variable, $\{\mathbf{I}^{(r,t)}, \forall r, t\}$. To reduce the time-complexity for solving the problem, we exploit the properties of the problem and develop a heuristic algorithm to obtain an approximate result. To simplify the notations, we eliminate the time index t in the remainder of the subsection since we focus on task partition and scheduling scheme for the tasks offloaded in one time slot. Let $r(z)$ and $h(z)$ represent the index of receiver and helper RSUs for task z, respectively.

Lemma 4.1 *If no task is queued after task z for both receiver RSU and helper RSU, the optimal partition ratio for the task is $x_z^* = \min\{\max\{0, \hat{x}_z\}, 1\}$, where $\hat{x}_z$ is determined by Eq. (4.16).*

$$\hat{x}_z = \begin{cases} \frac{X_1 + U_{z,h(z)}}{\chi W_z X_2}, & \text{if } X_1 \geq \chi X_2[W_z - R_{r(z),h(z)}(U_{z,h(z)} - A_z)] \\ \frac{(X_1 + A_z)R_{r(z),h(z)} + W_z}{\chi W_z(X_2 R_{r(z),h(z)} + 1)}, & \text{otherwise} \end{cases} \tag{4.16}$$

where

$$X_1 = \frac{\chi W_z}{C_{h(z)}} - \max\{U_{z,r(z)}, A_z\}, \tag{4.17}$$

and

$$X_2 = \frac{1}{C_{r(z)}} + \frac{1}{C_{h(z)}}. \tag{4.18}$$

Proof Without considering the service delay of tasks queued later, the optimal task partition ratio can be obtained by minimizing the following problem:

$$\min \max\{M_{z,r(z)}, M_{z,h(z)}\} \qquad \text{s.t. (4.14e).} \tag{4.19}$$

If $0 < x_z^* < 1$, the optimal task partition ratio can be obtained when $M_{z,r(z)} = M_{z,h(z)}$, i.e., $\hat{x}_z$. In addition, if $x_z^* = \max\{0, \hat{x}_z\} = 0$, helper RSU can fully process task z in a shorter service delay as compared to the queuing delay in receiver RSU, i.e., $X_1 \geq \max\{U_{z,h(z)}, A_z + \frac{\chi W_z}{R_{r(z),h(z)}}\}$. Otherwise, $x_z^* = \min\{1, \hat{x}_z\} = 1$, when receiver RSU can process task z by itself in a shorter service delay compared to the queuing delay in helper RSU, i.e., $\max\{U_{z,r(z)}, A_z\} \leq U_{z,h(z)} - \frac{\chi W_z}{C_{r(z)}}$. $\qquad \square$

Lemma 4.1 shows the optimal partition ratio from the individual task perspective. However, multiple tasks could be offloaded from different zones to an RSU, where the role of the RSU could be different for those tasks. The task partition strategy for a single task could affect the computing latency for the task queued later. Therefore, we will investigate the optimality of the task partition scheme in Lemma 4.1 in terms of minimizing the overall service delay for all tasks $z \in \mathcal{Z}$.

Lemma 4.2 *Assume that the following conditions are met:*

- *The computing capability, C_r, is identical for all edge servers.*
- *Receiver RSU and helper RSU are different for each task, i.e., $r(z) \neq h(z)$.*
- *For helper RSUs for all tasks, the queuing delay is not shorter than the offloading delay, i.e., $U_{z,h(z)} \geq A_z + F_{r(z),h(z)}, \forall z, r$.*

Then, given the execution order of tasks, the optimal solution of Problem (4.15) is given in Lemma 4.1, i.e., $x_z^* = \min\{\max\{0, \hat{x}_z\}, 1\}, \forall z$.

Proof An illustration of task partition is shown in Fig. 4.3. Given that task partition ratio following the results in Lemma 4.1, we focus on a task which is numbered as task 1 as shown in the figure. As indicated in the second and the third assumptions in Lemma 4.2, the computing load of task 1 is shared between RSU $r(1)$ and $h(1)$. Tasks 2 and 3 are scheduled after task 1 in RSU $r(1)$ and $h(1)$, respectively. In addition, $U_{2,h(2)} \geq A_2 + F_{r(2),h(2)}$. We then prove that, under the assumption in Lemma 4.2, the summation of the overall service delay of tasks 2 and 3 will increase if the partition ratio of task 1 does not follow the solution presented in Lemma 4.1.

For task 1, if the workload assigned to RSU $r(1)$ is decreased by Δx, the computing delay of task 1 in server $r(1)$ is reduced by $\Delta t_1^{r(1)} = \Delta x / C_{r(1)}$, while the computing delay of task 1 in server $h(1)$ is increased by $\Delta t_1^{h(1)} = \Delta x / C_{h(1)}$.

Fig. 4.3 An illustration of task partition

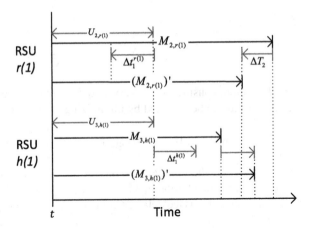

Correspondingly, the service delay of task 1 is increased by $\Delta T_1 = \Delta x / C_{h(1)}$. Given that task partition ratio x_1 is decreased by Δx, let $\hat{x}_2$ represent new partition ratio of task 2. Then, the service delay of task 2 can be calculated in the following cases:

- *Case 1:* RSU $r(1)$ is receiver RSU for task 2, i.e., $r(2) = r(1)$, and $\hat{x}_2 < 1$. According to Eq. (4.12) and Lemma 4.1, when the task partition ratio of task 1 is x_1, the service delay of task 2 is

$$
S_2 = \max\{A_2, U_{2,r(1)}\} + \frac{(U_{2,h(2)} - \max\{A_2, U_{2,r(1)}\})C_{h(2)} + \chi W_2}{C_{r(1)} + C_{h(2)}}.
$$
(4.20)

  After task partition ratio x_1 is decreased by Δx, task 2 can be processed by RSU $r(1)$ in advance by $\Delta t_1^{r(1)}$. The new service delay of task 2 is

$$
(S_2)' = \max\{A_2, U_{2,r(1)} - \Delta t_1^{r(1)}\}
$$
$$
+ \frac{(U_{2,h(2)} - \max\{A_2, U_{2,r(1)} - \Delta t_1^{r(1)}\})C_{h(2)} + \chi W_2}{C_{r(1)} + C_{h(2)}}.
$$
(4.21)

  The service delay deduction on task 2 can be obtained by subtracting Eq. (4.21) from Eq. (4.20). The deducted service delay for task 2 is $\Delta T_2 \leq \Delta t_1^{r(1)} C_{r(1)}/(C_{r(1)} + C_{h(2)})$.
- *Case 2:* RSU $r(1)$ is receiver RSU of task 2, i.e., $r(2) = r(1)$, and $\hat{x}_2 = 1$. When the task partition ratio of task 1 is x_1, the service delay of task 2 is

$$
(S_2)' = \max\{A_2, U_{2,r(1)} - \Delta t_1^{r(1)}\} + \frac{\chi W_2}{C_{r(1)}}.
$$
(4.22)

By subtracting Eq. (4.22) from Eq. (4.20), we have

$$\Delta T_2 \le \Delta t_1^{r(1)} - \frac{(\chi W_z/C_{r(1)} - U_{2,h(2)} + U_{2,r(1)})C_{h(2)}}{C_{r(1)} + C_{h(2)}}$$

$$\le \frac{\Delta t_1^{r(1)} C_{r(1)}}{(C_{r(1)} + C_{h(2)})} \tag{4.23}$$

where equality holds when $A_2 \le U_{2,r(1)} - \Delta t_1^{r(1)}$.

- Case 3: RSU $r(1)$ is helper RSU of task 2, i.e., $h(2) = r(1)$. When the task partition ratio of task 1 is x_1, the service delay of task 2 is

$$(S_2)' = \max\{A_2, U_{2,r(2)} - \Delta t_1^{r(2)}\}$$

$$+ \frac{(U_{2,r(1)} - \Delta t_1^{r(2)} - \max\{A_2, U_{2,r(2)}\})C_{r(1)} + \chi W_2}{C_{r(2)} + C_{r(1)}}. \tag{4.24}$$

Similar to Case 1, the deducted service delay for task 2 is $\Delta T_2 = \Delta t_1^{r(1)} C_{r(1)} / (C_{r(1)} + C_{r(2)})$.

As mentioned in the first assumption of Lemma 4.2, the computing capability, C_r, is identical for all servers. Thus, the maximum service delay deduction for task 2 is $\Delta t_1^{r(1)}/2$. For all tasks queued after task 1 in RSU $r(1)$, the overall service delay deduction is less than $\Delta t_1^{r(1)}[1/2 + (1/2)^2 + (1/2)^3 + \dots]$, which is always less than $\Delta t_1^{r(1)}$. The proof for the case when the workload assigned in RSU $h(1)$ is decreased by Δx can be obtained similarly. Therefore, the overall service delay will increase if the task partition ratio does not follow the solution presented in Lemma 4.2. □

Lemma 4.1 shows the optimal solution for the task partition ratio given the task execution order. Next, we will find the optimal scheduling order given the task partition ratio solution.

Lemma 4.3 *Consider only one RSU in the system, i.e., $r(z) = h(z), \forall z$. Assume that the offloading delay is proportional to the size of the task. The optimal task execution order is to process the task with the shortest service delay first.*

Proof Suppose that the tasks offloaded to edge server r are processed following the shortest-task-first rule, and task 2 is processed after the task 1. Then, we have

$$\max\{U_{1,r}, A_1\} + G_{1,r} \le \max\{U_{1,r}, A_2\} + G_{2,r}. \tag{4.25}$$

If task 1 and task 2 switch in the processing order, the service delay of task 2 decreases by

$$D = \max\{\max\{U_{1,r}, A_1\} + G_{1,r}, A_2\} - \max\{U_{1,r}, A_2\}. \tag{4.26}$$

On the other hand, the service delay of task 1 increases by

$$I = \max\{U_{1,r}, A_2\} + G_{2,r} - \max\{U_{1,r}, A_1\}. \tag{4.27}$$

From (4.25), we can derive that $I \geq G_{1,r}$. Then, the overall service delay of tasks 1 and 2 increases by

$$I - D \geq G_{1,r} - \max\{\max\{U_{1,r}, A_1\} + G_{1,r}, A_2\}$$
$$+ \max\{U_{1,r}, A_2\}. \tag{4.28}$$

We then consider the following three scenarios on A_2.

- *Case 1:* $A_2 \geq \max\{U_{1,r}, A_1\} + G_{1,r}$. In this case, $I - D \geq G_{1,r} \geq 0$.
- *Case 2:* $U_{1,r} \leq A_2 \leq \max\{U_{1,r}, A_1\} + G_{1,r}$. In this case, $I - D \geq A_2 - \max\{U_{1,r}, A_1\}$. According to the assumption in Lemma 4.3, where $A_1 \leq A_2$, we then have $I - D \geq 0$.
- *Case 3:* $A_2 \leq U_{1,r}$. In this case, $I - D \geq U_{1,r} - \max\{U_{1,r}, A_1\} = 0$.

Therefore, we can conclude that the overall service delay increases if the task execution order does not follow a shortest-task-first rule under the assumption in Lemma 4.3. □

Based on the properties provided in Lemmas 4.1–4.3, we design a heuristic algorithm, i.e., Algorithm 6, to determine the task execution order and allocate workload among the RSUs. In the algorithm, we allocate the task that has the shortest service delay first. For each task, receiver RSU and helper RSU share the workload according to the optimal partition ratio in Lemma 4.1. In the worst case, in which all zones have tasks to be offloaded in a time slot, the algorithm requires $|\mathcal{Z}|(|\mathcal{Z}|+1)/2$ iterations to compute the task partition and scheduling results, which can still provide fast response in a dynamic environment.

4.4 AI-Assisted Collaborative Computing Approach

We utilize a DRL method to solve the edge server selection problem in a dynamic environment. To implement the DRL method, we first re-formulate the problem into an MDP. An MDP can be defined by a tuple, $(\mathbb{S}, \mathbb{A}, \mathbb{T}, \mathbb{C})$, where $\mathbb{S}$ represents the set of system states; $\mathbb{A}$ represents the set of actions; $\mathbb{T} = \{p(s_{t+1}|s_t, a_t)\}$ is the set of transition probabilities; and $\mathbb{C}$ is the set of cost functions. Let $C(s, a)$ represent the cost when the system is at state $s \in \mathbb{S}$ and an action $a \in \mathbb{A}$ is taken. A policy, π, represents a mapping from $\mathbb{S}$ to $\mathbb{A}$. The state space, action space, and cost model in an MDP are summarized as follows:

(1) State space: In time slot t, the network state, s_t, includes the computing data amount in zones, i.e., $\{W_{z,t}, \forall z\}$, the average vehicle speed, i.e., $\{v_{z,t}, \forall z\}$, and

---

Algorithm 6 Task partition and scheduling algorithm (TPSA)

1: At time slot t, initialize set $\mathcal{S} = \{z | W_{z,t} \neq 0\}$.
2: Initialize $\psi_r = \hat{U}_{r,t}$, $\mathbf{I}^{(r,t)} = \mathbf{0}$, and $j_r = 1, \forall r$.
3: **while** $|\mathcal{S}| \neq 0$ **do**
4:     Initialize $Q_z = 0, \forall z \in \mathcal{S}$.
5:     **for** Task $z = 1 : |\mathcal{S}|$ **do**
6:         Update $r(z) = \{r | \alpha_{z,r,t} = 1\}$ and $h(z) = \{r | \beta_{z,r,t} = 1\}$.
7:         Update partition ratio $x_z = \min\{\max\{0, \hat{x}_z\}, 1\}$, where $\hat{x}$ is obtained by (4.16).
8:         Update $\hat{\psi}_{z,r(z)} = \psi_{r(z)} + M_{z,r(z)}$.
9:         Update $\hat{\psi}_{z,h(z)} = \psi_{h(z)} + M_{z,h(z)}$.
10:        If $x_z = 1$, then $Q_z = \hat{\psi}_{z,h(z)}$.
11:        If $x_z = 0$, then $Q_z = \hat{\psi}_{z,r(z)}$.
12:        If $0 < x_z < 1$, then $Q_z = (\hat{\psi}_{z,r(z)} + \hat{\psi}_{z,h(z)})/2$.
13:     **end for**
14:     Find $z^* = \text{argmin}_z Q_z$.
15:     Update $\psi_{r(z^*)} = \hat{\psi}_{z^*,r(z^*)}$ and $\psi_{h(z^*)} = \hat{\psi}_{z^*,h(z^*)}$.
16:     Update order matrix $I_{z^*,j_{r(z^*)}}^{r(z^*),t} = 1$, and $I_{z^*,j_{h(z^*)}}^{h(z^*),t} = 1$.
17:     Update $j_{r(z^*)} = j_{r(z^*)} + 1$, and $j_{h(z^*)} = j_{h(z^*)} + 1$.
18:     $\mathcal{S} = \mathcal{S} \backslash \{z^*\}$.
19: **end while**
20: $\hat{U}_{r,t+1} = \psi_r - \epsilon, \forall r$.

---

the delay for edge servers to finish computing the tasks offloaded in previous time slots $\{1, \ldots, t-1\}$, i.e., $\{\hat{U}_{r,t}, \forall r\}$.

(2) Action space: For zone z and time slot t, the action taken by the network includes three elements: the indices of receiver RSU, helper RSU, and deliver RSU, which can be represented by $\{a_{z,t}^1, a_{z,t}^2, a_{z,t}^3\}$, respectively.

(3) Cost model: Given a state-action pair, the overall service delay can be obtained. According to the objective function (4.14), the cost function can be formulated as

$$C(s_t, a_t) = \sum_{z \in \mathcal{Z}} \left\{ S_{z,t} \mathbf{1}_{z,t} + \lambda W_{z,t} (1 - \mathbf{1}_{z,t}) \right\}. \tag{4.29}$$

Then, the value function for the expected long-term discounted cost of state s is

$$V(s, \pi) = \mathbb{E}\left[\sum_{t=0}^{\infty} \gamma^t C(s_t, a_t) | s_0 = s, \pi \right] \tag{4.30}$$

where γ is a discount factor. By minimizing the value function of each state, we can obtain the optimal offloading and computing policy, π^*, which is

$$\pi^*(s) = \arg\min_a \sum_{s'} p(s'|s, a)[C(s, a) + \gamma V(s', \pi^*)]. \tag{4.31}$$

Due to the limited knowledge on the state transition probabilities and the sizeable state-action space in the network, the conventional dynamic programming technique is not able to find the optimal policy efficiently. Therefore, we adopt DRL to solve the server selection problem. There are three common DRL algorithms: deep Q network (DQN), actor-critic (AC), and DDPG. DQN is a powerful tool to obtain the optimal policy with a high dimension in the state space. DQN uses an online neural network (evaluation network) to learn the Q value and apply a frozen network (target network) to stabilize the learning process. However, the method shows inefficiency for problems with a high dimension in the action space, while the large number of zones leads to high dimensions in both state and action spaces in our problem. In comparison, both AC and DDPG can tackle problems with high dimensions in both state and action spaces by a policy gradient technique. Specifically, two networks, i.e., actor and critic networks, are adopted. The critic network evaluates the Q value, and the actor network updates policy parameters in the direction suggested by the critic. DDPG combines the characteristics of DQN on top of the AC algorithm to learn the Q value and the deterministic policy by adopting frozen networks, thereby achieving fast convergence [30]. In this chapter, we exploit the DDPG algorithm to obtain the optimal edge server selection policy in vehicular networks.

The illustration of our proposed AI-based collaborative computing approach is shown in Fig. 4.4. In each time slot, the controller observes the system state in the network. At state s_t, the optimal server selection policy can be generated by the DDPG algorithm. According to the server selection results, the corresponding task partition and scheduling policy can be obtained by the proposed TPSA algorithm. After server selection, task partition, and scheduling policies are deployed into the network, the cost of the corresponding state-action pair and the next system state are observed from the environment. A state transition tuple, (s_t, a_t, r_t, s_{t+1}), is stored in the replay memory for training neural networks. In DDPG, four neural networks are utilized. Two of the four networks are evaluation networks, where the weights keep updating whenever the neural networks are trained. The other two networks are target networks, where the weights are replaced periodically from evaluation networks. For both evaluation and target networks, two neural networks, i.e., actor and critic networks, evaluate the optimal policy and Q value, respectively. The weights in evaluation and target critic networks are denoted by θ^Q and $\theta^{Q'}$, and the weights in evaluation and target actor networks are denoted by θ^μ and $\theta^{\mu'}$, respectively.

In each training step, a batch of experience tuples are extracted from the experience replay memory, where the number of tuples in a mini-batch is denoted by N. Both evaluation and target critic networks determine the value function and compute loss function L, where

$$L(\theta^Q) = \mathbf{E}\left[\left(y_t - Q(s_t, a_t|\theta^Q)\right)^2\right]. \tag{4.32}$$

In (4.32), $Q(s_t, a_t|\theta^Q)$ represents the Q function approximated by the evaluation critic network. The value of y_t is obtained from the target critic network, where

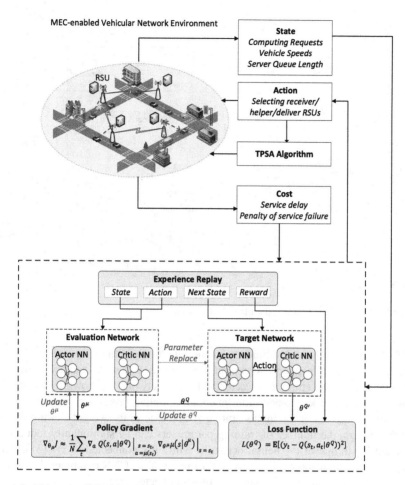

Fig. 4.4 AI-based collaborative computing approach

$$y_t = C(s_t, a_t) + \gamma \, Q(s_{t+1}, \mu'(s_{t+1}|\theta^{\mu'})|\theta^{Q'}). \tag{4.33}$$

In (4.33), $\mu'(s_{t+1}|\theta^{\mu'})$ represents the action taken at s_{t+1} given by the target actor network. On the one hand, the weights in the evaluation critic network, i.e., θ^Q, are updated by minimizing loss function (4.32). On the other hand, to update the weights of the evaluation actor network, the policy gradient can be represented as

$$\nabla_{\theta_\mu} J \approx \frac{1}{N} \sum_t \nabla_a Q(s, a|\theta^Q)|_{s=s_t, \atop a=\mu(s_t)} \nabla_{\theta^\mu} \mu(s|\theta^\mu)|_{s=s_t}. \tag{4.34}$$

According to (4.34), the weights of the evaluation actor network are updated in each training step.

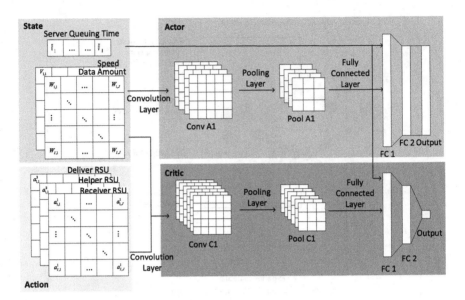

Fig. 4.5 The structure of actor and critic neural network

To further improve the efficiency of the DDPG algorithm, we utilize CNN in both actor and critic networks to exploit the correlation of states and actions among different zones. The structure of actor and critic networks is shown in Fig. 4.5. In those neural networks, convolution layers and pooling layers are applied to learn the relevant features of the inputs among zones. Due to the weight sharing feature of CNN filters, the number of training parameters can be significantly reduced as compared with that of the neural networks with fully connected layers [31]. After several convolution and pooling layers, the output of the CNN combines the state of edge servers and forwards the output to fully connected layers.

The proposed AI-based collaborative computing approach is implemented in Algorithm 7, where τ is a small number less than 1. As shown in Algorithm 7, to learn the environment efficiently, the system continuously trains the weights in the neural networks by N_t times after N_e time steps, where $N_e > N_t$.

4.5   Performance Evaluation

In this section, we first demonstrate the efficiency of the proposed TPSA algorithm in task partition and scheduling. Then, we evaluate the performance of the proposed AI-based collaborative computing approach in a vehicular network simulation using VISSIM [32].

Algorithm 7 AI-based collaborative computing approach

1: Initialize critic network $Q(s_0, a_0|\theta^Q)$ and actor network $\mu(s_0|\theta^\mu)$ with weights θ^Q and θ^μ.
2: Initialize target network with weights $\theta^{Q'} = \theta^Q$ and $\theta^{\mu'} = \theta^\mu$.
3: Initialize the experience replay buffer.
4: Initialize a random vector $\mathcal{N}$ as the noise for action exploration.
5: **for** episode = 1:G **do**
6:   Initialize environment, and observe the initial state s_0.
7:   **for** time slot $t = 1 : T$ **do**
8:     Select action $a_t = \mu(s|\theta^\mu) + \mathcal{N}$.
9:     Let $\alpha_{z,a_{z,t}^1,t}$, $\beta_{z,a_{z,t}^2,t}$, and $\gamma_{z,a_{z,t}^3,t}$ equal to 1.
10:     Compute the task partition and scheduling results by Algorithm 6
11:     Observe next state s_{t+1} and cost $C(s_t, a_t)$.
12:     Store transition (s_t, a_t, r_t, s_{t+1}) into the experience replay buffer. Delete the oldest transition set if the buffer is full.
13:     **if** $k \mod N_e == 0$ **then**
14:       **for** $j = 1 : N_t$ **do**
15:         Sample a mini-batch of N samples.
16:         Update y_t by (4.33).
17:         Update the weights in the evaluation critic network by minimizing the loss in (4.32).
18:         Update the weights in the evaluation actor network using sampled policy gradient presented in (4.34).
19:         Update target networks: $\theta^{Q'} = \tau\theta^Q + (1 - \tau)\theta^{Q'}; \theta^{\mu'} = \tau\theta^\mu + (1 - \tau)\theta^{\mu'}$.
20:       **end for**
21:     **end if**
22:   **end for**
23: **end for**

4.5.1 Task Partition and Scheduling Algorithm

We first evaluate the performance of the proposed TPSA algorithm. In the simulation, computing tasks can be offloaded to five RSUs with the offloading rate of 6 Mbits/s. The communication rate among the servers is 8 Mbits/s. The computing capability of the servers is 8 Gigacycle/s, and the number of computing cycles needed for processing 1 Mbit is 4 Gigacycle . The amount of data for each computing task is uniformly distributed between 1 Mbits and 21 Mbits. For each task, receiver and helper RSUs are randomly selected from the five RSUs. We compare the proposed TPSA algorithm with *brute-force* and *random* schemes. In the brute-force scheme, the exhaustive search is utilized to find the optimal scheduling order. In the random scheme, the task execution order is randomly assigned. Note that, for both *brute-force* and *random* schemes, we apply the optimal task partition ratio in Lemma 4.2. The simulation results are averaged over 200 rounds of Monte Carlo simulations.

The service delay performance is shown in Fig. 4.6a. It can be seen that, as the task number increases, the overall service delay increases correspondingly, and the increasing rate of the random scheme is the highest among the three schemes. The proposed TPSA algorithm achieves a performance very close to the brute-force scheme. Moreover, we compare the runtime between the proposed TPSA and the

Fig. 4.6 (**a**) Average service delay among the three task partition and scheduling schemes with respect to the number of tasks. (**b**) Average computing runtime among the three task partition and scheduling schemes with respect to the number of tasks

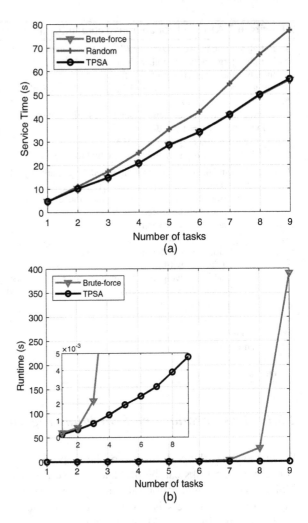

brute-force scheme. As shown in Fig. 4.6b, when the number of task increases, the runtime of brute-force scheme increases exponentially, while the proposed TPSA algorithm has insignificant runtime to determine the task execution order. In summary, the proposed TPSA algorithm can achieve a near-optimal performance for task partition and scheduling with low computing complexity.

4.5.2 AI-Based Collaborative Computing Approach

In this subsection, we evaluate the performance of the proposed AI-based collaborative computing approach. In the simulation, we consider an 800 m × 800 m

Fig. 4.7 The transportation network topology for simulation

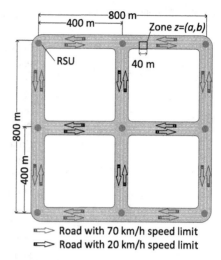

Table 4.1 Network parameters

P^V	P^R	σ_r^2, σ_v^2	λ	ϵ
27 dBm	37 dBm	−93 dBm	50	1 s
f	χ	N_e, N_t	δ^O	δ^D
2800 MHz	1200 C/bits	80, 25	7 dB	7 dB

transportation system, where the transportation topology is shown in Fig. 4.7. Nine RSUs with edge servers are deployed, as shown in the figure. We generate vehicle traffic by VISSIM [32], with 200 vehicles traveling in the area. The speed of vehicles depends on the speed limit of the road and the distance between vehicles. For each vehicle, the computing tasks are generated following a Poisson process, and the input data amount of a task is uniformly distributed in the range of [2,5] Mbits. The length and width of a zone are 40 m and 10 m (2 driving lanes), respectively. Other network parameter setting is presented in Table 4.1. We simulate the system performance within a duration of 20 s.

The neural network structure of the DDPG algorithm is presented in Table 4.2. The initial learning rates of the actor and critic networks are 1e-5 and 1e-4, respectively, and the learning rates are attenuated by 0.991 in every 500 training steps. The experience replay buffer can adopt 8000 state-action transition tuples. In each training step, the number of transition tuples selected for training, i.e., the batch size, is 128. We adopt a soft parameter replacement technique to update the weights in target networks, where τ is 0.01. We compare the performance of the proposed AI-based collaborative computing approach with three benchmark approaches. In the Greedy approach, vehicles always offload their tasks to the RSU with the highest SNR, and receiver RSU does not share the workload with other RSUs. In the Greedy+TPSA approach, a vehicle offloads its tasks to the RSU with the highest SNR, and receiver RSU randomly selects another RSU to compute the task collaboratively. The task partition and scheduling policy follows the proposed

Table 4.2 Neural network structure

Actor network		
Layer	Number of neurons	Activation function
CONV1	$5 \times 1 \times 2 \times 10$, stride 1	relu
POOL1	2×1	none
Data concatenation and batch normalization layer		
FC1	1400	tanh
FC2	1400	tanh
FC3	$5 \times A \times B$	tanh
Critic network		
Layer	Number of neurons	Activation function
CONV1	$5 \times 1 \times 2 \times 40$, stride 1	relu
POOL1	2×1	none
CONV2	$3 \times 1 \times 40 \times 10$, stride 1	relu
POOL2	2×1	none
Data concatenation and batch normalization layer		
FC1	640	relu
FC2	512	relu
FC3	128	none
FC4	1	relu

Fig. 4.8 Average weighted computing service cost versus computing task arrival rate per vehicle

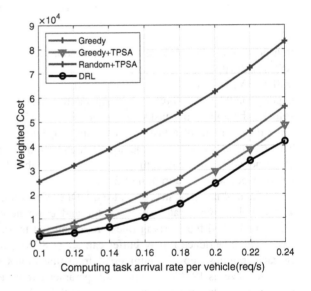

TPSA algorithm, and computing results are delivered by the receiver RSU. In the Random+TPSA approach, receiver, helper, and deliver RSUs are selected randomly, and the TPSA algorithm is applied to determine the task partition ratio and the execution order.

The overall weighted service cost with respect to task arrival rate is shown in Fig. 4.8. Our proposed approach can achieve the lowest computing cost compared

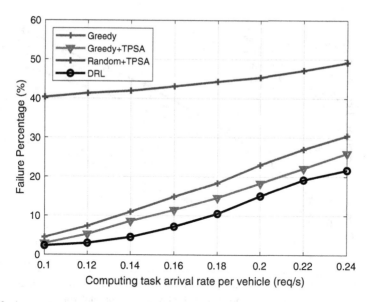

Fig. 4.9 Average percentage of service failure versus computing task arrival rate per vehicle

with the other three approaches. The reason is that parallel computing is able to reduce the overall service delay, and the proposed TPSA is able to achieve near-optimal task partition and scheduling results. The random approach suffers the highest cost compared with others due to inefficient server selection. The Greedy+TPSA approach achieves a lower cost than the greedy approach. However, both Greedy+TPSA and greedy approaches select servers according to the instantaneous cost rather than the value in the long term. Therefore, the Greedy+TPSA approach cannot attain a lower cost as compared to the proposed AI-based approach.

As indicated in Eq. (4.29), the weighted service cost consists of the service delay and the failure penalty. The percentage of service failure is shown in Fig. 4.9. Similar to the service cost, the proposed AI-based approach achieves the lowest failure percentage among the four approaches. Correspondingly, as shown in Fig. 4.10, the proposed approach can successfully process the highest amount of data among the four approaches. The average service delay for successfully computing 1 Mbits data is shown in Fig. 4.11. Compared with the other three approaches, the proposed approach reduces the service delay significantly. Furthermore, the delay of the random approach increases exponentially since less data can be successfully computed when the task arrival rate is high.

The convergence performance of the proposed AI-based approach is shown in Fig. 4.12, where the highlighted line represents the moving average over 50 samples around the corresponding point. Note that our algorithm explores the environment multiple times in each training step. It can be seen that our approach converges after 10,000 episodes, or equivalently, after the network is trained by around 3000 episodes, i.e., 60,000 training steps.

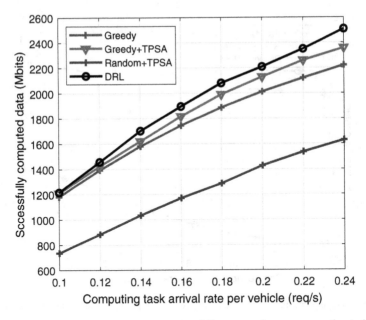

Fig. 4.10 Average computing data amount successfully computed versus computing task arrival rate per vehicle

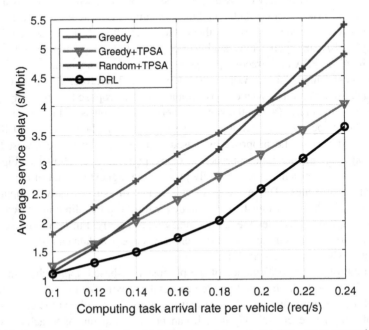

Fig. 4.11 Average service delay for 1 Mbits successful computed data versus computing task arrival rate per vehicle

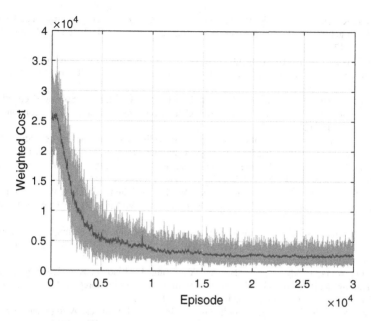

Fig. 4.12 Convergence performance of the proposed algorithm, where the task arrival rate is 0.1 request/sec

4.6 Summary

We have introduced a novel collaboration computing framework to reduce computing service latency and improve service reliability in MEC-enabled vehicular networks. The proposed framework addresses the challenge of maintaining computing service continuity for vehicles with high mobility. As a result, our collaborative computing approach is able to support proactive decision making for computing offloading through learning the network dynamics. Our work can be applied to offer low-latency and high-reliability edge computing services to vehicles in a complex network environment, such as urban transportation systems. A potential future direction is using a multi-agent learning approach to compute the optimal computing strategy with limited information collected by the edge servers.

References

1. Peng, H., Shen, X.: Deep reinforcement learning based resource management for multi-access edge computing in vehicular networks. IEEE Trans. Netw. Sci. Eng. **7**(4), 2416–2428 (2020)
2. Gao, J., Li, M., Zhao, L., Shen, X.: Contention intensity based distributed coordination for V2V safety message broadcast. IEEE Trans. Veh. Technol. **67**(12), 12,288–12,301 (2018)
3. Self Driving Safety Report: Tech. rep., Nvidia, Santa Clara, CA, USA (2018)

4. Zhang, K., Zhu, Y., Leng, S., He, Y., Maharjan, S., Zhang, Y.: Deep learning empowered task offloading for mobile edge computing in urban informatics. IEEE Internet Things J. **6**(5), 7635–7647 (2019)
5. Liu, J., Wan, J., Zeng, B., Wang, Q., Song, H., Qiu, M.: A scalable and quick-response software defined vehicular network assisted by mobile edge computing. IEEE Commun. Mag. **55**(7), 94–100 (2017)
6. Zhang, N., Zhang, S., Yang, P., Alhussein, O., Zhuang, W., Shen, X.: Software defined space-air-ground integrated vehicular networks: Challenges and solutions. IEEE Commun. Mag. **55**(7), 101–109 (2017)
7. Xu, J., Chen, L., Ren, S.: Online learning for offloading and autoscaling in energy harvesting mobile edge computing. IEEE Trans. Cogn. Commun. Netw. **3**(3), 361–373 (2017)
8. Zhang, J., Letaief, K.B.: Mobile edge intelligence and computing for the Internet of Vehicles. Proc. IEEE **108**(2), 246–261 (2020)
9. Lyu, F., Zhu, H., Zhou, H., Qian, L., Xu, W., Li, M., Shen, X.: MoMAC: Mobility-aware and collision-avoidance MAC for safety applications in VANETs. IEEE Trans. Veh. Technol. **67**(11), 10,590–10,602 (2018)
10. He, Y., Zhao, N., Yin, H.: Integrated networking, caching, and computing for connected vehicles: A deep reinforcement learning approach. IEEE Trans. Veh. Technol. **67**(1), 44–55 (2018)
11. Qi, Q., Wang, J., Ma, Z., Sun, H., Cao, Y., Zhang, L., Liao, J.: Knowledge-driven service offloading decision for vehicular edge computing: A deep reinforcement learning approach. IEEE Trans. Veh. Technol. **68**(5), 4192–4203 (2019)
12. Wang, X., Ning, Z., Wang, L.: Offloading in Internet of Vehicles: A fog-enabled real-time traffic management system. IEEE Trans. Ind. Inf. **14**(10), 4568–4578 (2018)
13. Sun, Y., Guo, X., Song, J., Zhou, S., Jiang, Z., Liu, X., Niu, Z.: Adaptive learning-based task offloading for vehicular edge computing systems. IEEE Trans. Veh. Technol. **68**(4), 3061–3074 (2019)
14. Ning, Z., Zhang, K., Wang, X., Obaidat, M.S., Guo, L., Hu, X., Hu, B., Guo, Y., Sadoun, B., Kwok, R.Y.K.: Joint computing and caching in 5G-envisioned Internet of Vehicles: A deep reinforcement learning-based traffic control system. IEEE Trans. Intell. Transp. Syst. **22**(8), 5201–5212 (2021)
15. Hafeez, K.A., Zhao, L., Liao, Z., Ma, B.N.: A fuzzy-logic-based cluster head selection algorithm in VANETs. In: 2012 IEEE Int. Conf. Commun. (ICC), pp. 203–207 (2012)
16. Taleb, T., Ksentini, A., Frangoudis, P.A.: Follow-me cloud: When cloud services follow mobile users. IEEE Trans. Cloud Comput. **7**(2), 369–382 (2019)
17. Taleb, T., Ksentini, A.: Follow me cloud: interworking federated clouds and distributed mobile networks. IEEE Netw. **27**(5), 12–19 (2013)
18. Urgaonkar, R., Wang, S., He, T., Zafer, M., Chan, K., Leung, K.: Dynamic service migration and workload scheduling in edge-clouds. Perform. Eval. **91**, 205 – 228 (2015)
19. Wang, S., Urgaonkar, R., Zafer, M., He, T., Chan, K., Leung, K.K.: Dynamic service migration in mobile edge computing based on Markov decision process. IEEE/ACM Trans. Netw. **27**(3), 1272–1288 (2019)
20. Liang, H., Zhang, X., Zhang, J., Li, Q., Zhou, S., Zhao, L.: A novel adaptive resource allocation model based on SMDP and reinforcement learning algorithm in vehicular cloud system. IEEE Trans. Veh. Technol. **68**(10), 10,018–10,029 (2019)
21. Mao, Y., You, C., Zhang, J., Huang, K., Letaief, K.B.: A survey on mobile edge computing: The communication perspective. IEEE Commun. Surv. Tuts. **19**(4), 2322–2358 (2017)
22. Cao, J., Yang, L., Cao, J.: Revisiting computation partitioning in future 5G-based edge computing environments. IEEE Internet Things J. **6**(2), 2427–2438 (2019)
23. Lin, L., Liao, X., Jin, H., Li, P.: Computation offloading toward edge computing. Proc. IEEE **107**(8), 1584–1607 (2019)
24. Zhou, J., Tian, D., Wang, Y., Sheng, Z., Duan, X., Leung, V.C.M.: Reliability-optimal cooperative communication and computing in connected vehicle systems. IEEE Trans. Mobile Comput. **19**(5), 1216–1232 (2020)

25. Xiao, Z., Dai, X., Jiang, H., Wang, D., Chen, H., Yang, L., Zeng, F.: Vehicular task offloading via heat-aware MEC cooperation using game-theoretic method. IEEE Internet Things J. **7**(3), 2038–2052 (2020)
26. Balalaie, A., Heydarnoori, A., Jamshidi, P.: Microservices architecture enables DevOps: Migration to a cloud-native architecture. IEEE Softw. **33**(3), 42–52 (2016)
27. Kuang, Z., Li, L., Gao, J., Zhao, L., Liu, A.: Partial offloading scheduling and power allocation for mobile edge computing systems. IEEE Internet Things J. **6**(4), 6774–6785 (2019)
28. Li, M., Cheng, N., Gao, J., Wang, Y., Zhao, L., Shen, X.: Energy-efficient UAV-assisted mobile edge computing: Resource allocation and trajectory optimization. IEEE Trans. Veh. Technol. **69**(3), 3424–3438 (2020)
29. Chen, J., Xu, W., Cheng, N., Wu, H., Zhang, S., Shen, X.: Reinforcement learning policy for adaptive edge caching in heterogeneous vehicular network. In: 2018 IEEE Global Commun. Conf. (GLOBECOM), pp. 1–6 (2018)
30. Lillicrap, T.P., Hunt, J.J., Pritzel, A., Heess, N., Erez, T., Tassa, Y., Silver, D., Wierstra, D.: Continuous control with deep reinforcement learning. Preprint (2015). arXiv:1509.02971
31. Krizhevsky, A., Sutskever, I., Hinton, G.E.: ImageNet classification with deep convolutional neural networks. In: Advances in Neural Information Processing Systems, vol. 25, pp. 1097–1105 (2012)
32. VISSIM. http://vision-traffic.ptvgroup.com/

Chapter 5
Edge-Assisted Mobile VR

In this chapter, we investigate edge-assisted content caching and distribution for mobile virtual reality (VR) video streaming. In the considered VR use case, an edge server caches chunks of popular videos and delivers them to users. We focus on how a caching policy can improve the quality of experience of users while adapting to network dynamics given limited network resources. The research objective is to develop content placement and distribution schemes to cache high-quality and popular VR video chunks in a video-on-demand setting while reducing the video frame missing rate. To achieve this objective, we first propose a content placement scheme to make decisions on which video chunks to cache, while considering the trade-off among communication, computing, and caching resource utilization. Based on the content placement decisions, we further propose a learning-based content distribution scheme to allocate computing units given content delivery requests from multiple users. Specifically, a deep reinforcement learning approach is developed for the edge server to deliver content effectively and reduce the overall frame missing rate with low communication overhead. Simulation results are provided to verify the performance of proposed schemes.

5.1 Background on Mobile Virtual Reality

Mobile VR streaming is an emerging application of importance in next-generation wireless networks [1]. In the scenario of mobile VR, users can watch 360° VR videos using wireless head-mounted devices (HMDs). A key feature of mobile VR videos is their ultra-high spatial resolution. A mobile VR video can have a resolution up to 12K (11,520 × 6480 pixels), while conventional videos normally have a resolution of 4K or less [2]. From the communication perspective, delivering VR videos to users requires an unprecedented high bit rate, which introduces pressure on both backhaul and wireless links in existing communication networks. From the

© The Author(s), under exclusive license to Springer Nature Switzerland AG 2021
J. Gao et al., *Connectivity and Edge Computing in IoT: Customized Designs and AI-based Solutions*, Wireless Networks,
https://doi.org/10.1007/978-3-030-88743-8_5

computing perspective, extensive video processing and frame rendering require high computational capability, while the computing capability of the GPUs in HMDs is too limited to support an acceptable frame rate and render videos with satisfactory latency, even for a modest video quality [3].

MEC leverages computing and storage resources on network edge, such as small BSs and other APs, to cache popular content and process compute-intensive tasks offloaded by users. Specifically, the storage resource at an edge server can be utilized to cache popular VR video chunks to alleviate network congestion in the backbone, and the computing resource can be utilized to process video chunks for HMDs [4, 5]. However, there are many challenges in implementing MEC-assisted mobile VR applications. Even for one-hop communications between edge servers and VR devices, i.e., HMDs, it is difficult to achieve an ultra-high bit rate for VR video delivery. In particular, VR users will feel dizzy if the motion-to-photon latency, i.e., the time elapsed between user movement and VR display response, is greater than 20 ms [6]. For a video segment of 12K resolution with a duration of two seconds and 60 frames per second, content delivery would have to reach over 3000 Gbps to meet the stringent motion-to-photon latency requirement, which is too high for the current communication networks. Furthermore, since the storage resource at the edge server is limited, it is impossible to store all VR videos at the edge server. Therefore, some VR videos have to be downloaded from the cloud server with a higher content delivery delay, which further increases the difficulty of satisfying the requirements of mobile VR video delivery. Thus, to support VR video streaming, innovations in video caching and delivery methods are necessary.

Innovative content delivery should take advantage of VR video characteristics and the layered encoding technique to reduce the resource consumption for delivering VR videos. Since VR videos are 360° panoramic, a VR user only watches a part of the spherical domain of a video, referred to as a field-of-view (FoV), at any instant. Therefore, a full VR video can be divided into small spatio-temporal chunks, and only the chunks corresponding to the user's current FoV are delivered to HMDs. Due to the comparatively small size of a video chunk, the resolution of the video to be delivered can be reduced significantly, which results in a lower data rate requirement for video delivery. In addition, facilitated by AI techniques, the watching preferences and motions of users can be predicted in advance. Using the prediction, video chunks can be delivered to the HMDs proactively and stored in the buffer of the devices, which increases the tolerance for content delivery latency. The above two solutions make mobile VR video streaming possible, while proper resource management schemes are needed to implement them.

5.2 Caching and Computing Requirements of Mobile VR

There are three aspects of FoV-based content caching and proactive content delivery to investigate. First, the format of VR video chunks to be cached and delivered to users should be determined. Different video chunk sizes result in different video encoding efficiency and file sizes, which further impact the efficiency of content caching and delivery. Second, a caching policy should be developed for resource-efficient edge caching. Finally, given cached video chunks, it is necessary to schedule limited computing units at the edge server in order to enable proactive video processing and content delivery. In this section, we investigate how to address the caching and computing requirements in mobile VR video delivery from these aspects.

5.2.1 Mobile VR Video Formats

As aforementioned, a VR video can be divided into small video chunks, and only video chunks in the user's FoV need to be delivered. While this method reduces the file size in video delivery, it raises the problem of granularity while dividing video chunks. Multiple video chunks with a small size can be delivered to an HMD to render an FoV video [2, 7, 8]. Fine video quality can be achieved by utilizing different video qualities in different video chunks. However, dividing videos into small-size chunks leads to low video encoding efficiency in storing the chunks and high complexity for selecting video chunks from a large number of candidates [9]. Alternatively, one video chunk with a large size can be delivered to an HMD to render the FoV video [5, 10]. However, video chunks with a large size require more communication resources in delivery, and the quality of an FoV video cannot be adjusted flexibly to adapt to network dynamics. Therefore, the size of video chunks should be properly selected to balance the video encoding efficiency and the granularity of video quality adjustment.

5.2.2 Edge Caching for Mobile VR

Popular video chucks can be prefetched to reduce content delivery latency by using the caching resource of an edge server. However, compared to conventional video caching, determining the popularity of video chunks in VR videos is more challenging. In conventional videos, video segments (VSs) are divided temporally, and the popularity of VSs in one video is similar and follows the popularity of the corresponding video. However, for mobile VR, video chunks are divided spatio-temporally, and the popularity of video chunks is determined by the movement of the user's viewpoint. In order to determine the popularity of video chunks, extensive measurements must be made by HMDs during video playback [11]. Fine-grained measurement of popularity complicates edge caching, since a video could have millions of video chunks that may have different popularity. Moreover, with small-size video chunks, rendering a user FoV needs multiple video chunks; with large-size video chunks, a video chunk may render multiple FoVs. As a result, correlation on popularity may exist among video chunks, which makes edge caching more difficult. Without knowledge of the spatial correlations among video chunks, the effectiveness of a caching policy cannot be guaranteed. Therefore, to support mobile VR video delivery, a proper content placement scheme should evaluate the popularity of video chunks in a scalable manner according to their spatial correlations.

Furthermore, different from conventional caching, the edge server can cache both original video chunks and processed video chunks in mobile VR video caching. The resulting trade-off among communication, caching, and computing in mobile VR video caching has been discussed in [10, 12], in which caching solutions are based on a given popularity profile of video chunks and a constant processing rate. An edge server with a high computing capability should prefetch more original video chunks in the cache to avoid frequent content fetching from the cloud server, while an edge server with a low computing capability should prefetch more processed video chunks to reduce processing delay. However, due to the dynamics of content delivery requests and limited computing units at the edge server, the content delivery delay dynamically changes with the content delivery demands. As a result, a constant processing rate without considering dynamically changing content delivery demands may not be accurate for evaluating the content placement policy. To optimize the content delivery policy, the dynamics of content delivery demand should be evaluated while investigating the trade-off among multi-dimensional resources.

5.2.3 Edge Computing for Mobile VR

Given a content placement policy, how to properly deliver cached VR video chunks to users is another challenge. Although the average content delivery latency may be reduced by caching popular video chunks, satisfying the real-time delivery latency is difficult, especially when unpopular video chunks are requested. Unsuccessful content delivery results in failure of rendering video frames i.e., frame missing, at user HMDs. Proactive video chunk delivery and scheduling policies based on predicted user HMD trajectory or user request profile have been considered in [13, 14] to alleviate frame missing. However, the impact of network dynamics on the scheduling policy should be further investigated.

Furthermore, proactive content delivery is based on prediction. In particular, an HMD can track the user's viewpoint trajectory, predict viewpoints in the subsequent time slots, and request the corresponding video chunks to render the FoV of the predicted viewpoints [15]. The accuracy of viewpoint prediction can therefore have a significant impact on content delivery. Various learning-based techniques have been adopted to improve the accuracy of viewpoint prediction, such as long short-term memory (LSTM) networks [15, 16] and linear regression methods [17]. However, it is difficult to achieve error-free prediction. To improve the quality of experience of users, such uncertainty should be taken into account in proactive content delivery.

In the remainder of this chapter, we investigate the problem of edge-assisted content caching and delivery for mobile VR video streaming. The research objective is to develop content placement and distribution schemes to cache popular and high-quality VR video chunks and deliver video chunks to users while reducing video frame missing. Specifically, we focus on three research problems: determining resource-efficient video chunk formats for caching, designing content placement policy subject to network resource constraints, and developing a content distribution scheme to accommodate network and user request dynamics. To enable a scalable and efficient manner for content placement for mobile VR video streaming, we look into the specific characteristics of VR videos, evaluate the trade-off among multi-dimensional resources, and propose a content placement solution to handle the complex problem with a large number of variables. Meanwhile, for content distribution, we explore a novel learning-based solution to schedule content delivery for users with low communication overhead.

5.3 Mobile VR Video Caching and Delivery Model

In this section, we elaborate on the system model in mobile VR video caching and delivery and present the research problems given the system model.

5.3.1 Network Model

We consider a network with multiple VR users with HMDs within the communication coverage of an edge server, such as a BS or an AP, as illustrated in Fig. 5.1. A wired communication link connects the cloud server on the Internet and the edge server, and the average data rate is R_B on the wired link. The edge server communicates with user HMDs through wireless communication links. The overall number of users is denoted by U.

The edge server is equipped with storage and computing capabilities for content caching and processing, respectively. Popular content can be prefetched at the cache for reducing traffic on the wired link and alleviating content delivery latency. The cache capacity of the edge server is denoted by C. When a video chunk is requested by a user HMD, the video chunk can be downloaded from the edge server, if the corresponding content is cached, or from the cloud server otherwise. User HMDs play 360-degree 3D stereoscopic videos, where extensive computing is required for video processing, such as projection between 2D monoscopic to 3D stereoscopic videos. Since the computing capability at a user HMD is limited, most video processing is executed at the edge server. In an edge server, there are E computing units, and the computing frequency for video processing is f. The computing model is introduced in Sect. 5.3.2.

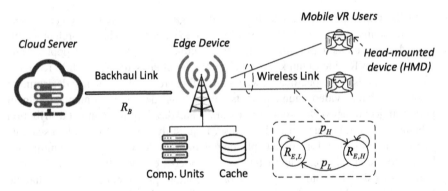

Fig. 5.1 An illustration of the network model

The edge server connects with a user HMD at a mmWave band, while a backup band, such as a sub-6GHz band, is utilized when the high-speed mmWave band is in outage [18]. Such transmission link is modeled as a two-stage Markov chain. When the high-speed mmWave band is in operation, the corresponding transmission rate is $R_{E,H}$. Otherwise, when the backup band is in operation, the corresponding transmission rate is $R_{E,L}$. The probabilities of the high-speed state (with rate $R_{E,H}$) transiting from and to the low-speed state (with rate $R_{E,L}$) are denoted by p_L and p_H, respectively. Independent Markov processes among user HMDs are assumed. The corresponding average transmission rate between the edge server and a user HMD is denoted by $\bar{R}_E$.

5.3.2 Content Distribution Model

Similar to a conventional video, a full-length VR video is divided into VSs in the *time domain* for video streaming, which is illustrated as the horizontal layers in Fig. 5.2. The playback time duration for a VS is T_S. The group of VSs corresponding to video l is denoted by S_l. As aforementioned, a VR user watches a part of the spherical domain of a video, i.e., an FoV, at any given instant. The center point of an FoV is called the viewpoint, which changes with the user HMD movement. Delivering full 360° video is bandwidth-consuming, and the typical solution is to deliver only video within the user's FoV. A VS is further divided into *video chunks* in the *spatial domain*. The minimum unit of the spatial area of a chunk is referred to as a *tile*. Each VS is evenly divided into $X \times Y$ tiles in the spatial domain, shown as the cubes in Fig. 5.2. The set of tiles is denoted by $\mathcal{I}$, and $|\mathcal{I}| = X \times Y$. A video chunk may consist of several tiles, and one or multiple video chunks are assembled to render an FoV. Details of video chunk formats are given in Sect. 5.4.1.

A typical system model for 360° stereoscopic VR delivery is shown in Fig. 5.3. The cloud server stores 360° equirectangular videos in 2D with auxiliary 3D representation files, such as depth maps [20]. With the 3D representation files, 2D

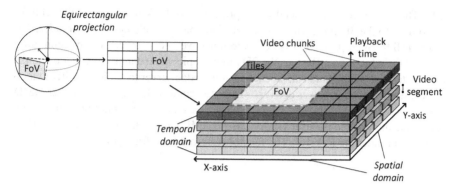

Fig. 5.2 VR video structure and division

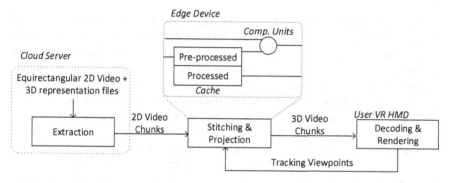

Fig. 5.3 VR video delivery model [10, 19]

video chunks can be projected to 3D video chunks. We do not investigate the caching policy for the 3D representation files in this work since they have much smaller data size in comparison with video chunks. The video delivery is based on layered encoding, where a video is encoded into a base layer, which has a low resolution to satisfy the minimum video quality requirement of users, and an enhanced layer, which can be added on top of the base layer for video quality improvement. To reduce the bandwidth consumption in video delivery, the equirectangular video is further divided into video chunks both spatial and temporal domains. The popular 2D video chunks can be prefetched to the cache at the edge server.

A user HMD continuously tracks the viewpoint movement, with the viewpoint sampling frequency g. When a viewpoint of the user changes, a corresponding 3D stereoscopic video chunk may be requested from the edge server. The edge server can download the video chunk from the cloud server or fetch it from the cache at the edge server. As two different videos are required for the left and the right eyes of each user, the data size of a stereoscopic video chuck is larger than the data size of the corresponding monoscopic video chunk [10, 21]. Therefore, to alleviate the backhaul traffic load, the video chunks downloaded from the cloud server are in

2D. The 2D video chunks need to be projected to 3D stereoscopic video chunks according to the 3D representation files before the video chunks can be rendered at user HMDs. Such projection is compute-intensive and is processed by computing units at the edge server. Besides the 2D monoscopic video chunks, the video chunks in 3D stereoscopic format can be cached at the edge server to avoid repetitive video processing. The edge server delivers 3D stereoscopic video chunks to a user HMD for further decoding and rendering. If the requested content cannot be delivered in time or cannot match the user viewpoint, the HMD can render only a part of its FoV or cannot render the FoV [6]. The result is referred to as frame missing.

5.3.3   Content Popularity Model

We investigate proactive caching at the edge server, which utilizes statistical popularity information and prefetches popular video chunks. The popularity of VS s is denoted by p_s^{VS}. User viewpoints may focus on a similar region when they are watching a specific VS, i.e., region of interest. The region of interest highly depends on the video content. The averaged fraction of time that user viewpoints fall into a specific tile over the whole duration of a VS is referred to as the viewpoint popularity, which can be obtained from historical user viewpoint movement profiles created during VR streaming [11]. User HMDs track the user viewpoint movement, and the edge server and the corresponding service provider can collect data from HMDs to obtain the viewpoint distribution for each VS. The viewpoint popularity distribution for VS s is denoted by $\mathbf{p}_s^P = [p_{1,s}^P, \ldots, p_{|\mathcal{I}|,s}^P]$, where $p_{i,s}^P$ represents the viewpoint popularity at tile i in VS s. The popularity of video chunks can be obtained by viewpoint popularity distributions. As a result, the edge server can proactively cache video chunks according to their popularity. In our work, the popularity distributions of VSs and viewpoints are assumed to be known in advance.

5.3.4   Research Objective

Given the system model, the research problem is to design content caching and distribution schemes for mobile VR video streaming with the following three objectives:

- Caching the popular and high-quality VR video chunks at the edge server to reduce backhaul traffic, subject to average content delivery time and cache capacity constraints;
- Reducing video streaming frame missing likelihood to improve the quality of experience of mobile VR users within the coverage of an edge server;
- Exploring the role of learning methods in the content caching and distribution scheme.

In addition, we have the following design targets:

- Improving the resource utilization efficiency on content caching and distribution;
- Reducing the message overhead for real-time content distribution.

Caching mobile VR videos has two differences from caching conventional videos: diversified content formats and an even larger amount of content. The differences result in different content caching considerations. In terms of content formats, both 2D monoscopic video chucks and 3D stereoscopic video chunks can be cached at the edge. Although 2D monoscopic video chunks have a smaller data size, caching 2D video chunks yields computing latency when the data is delivered to the users. Caching 3D stereoscopic video chunks can reduce the computing latency, while they require more storage resources. A trade-off among caching, computing, and communication resources should be carefully made in the caching policy. In terms of the scale of content, a VR video can be divided into video chunks into both temporal and spatial dimensions. As a result, the number of video chunks associated with a VR video can increase significantly from that associated with a traditional video. A scalable cache management policy is required, which should address the design challenges of the diversified popularity distribution and content formats. Furthermore, even in conventional video streaming, content distribution is challenging in a dynamic network environment, especially when multiple users are served simultaneously by an edge server. Although edge caching reduces the content delivery latency for popular content, large delivery latency can occur when unpopular content is requested and delivered. For VR videos, the user viewpoint movement further introduces uncertainty in content distribution. In such a dynamic environment, video delivery scheduling without yielding extra overhead is difficult.

To achieve the research objectives, we first design an adaptive model for dividing video chunks in Sect. 5.4.1. We then propose a content placement scheme to cache high-quality and popular video chunks at the edge server subject to network constraints in Sect. 5.4.2. Furthermore, a proactive content distribution scheme is proposed to improve the content delivery performance in real-time in Sect. 5.5.

5.4 Content Caching for Mobile VR

In this section, we propose a novel content placement scheme for caching mobile VR video chunks. It achieves a balance between the popularity of cached video chunks and the video quality, while satisfying the resource constraints on communication, computing, and caching. By evaluating the trade-off among network resources, an edge server can adapt its caching policy to alleviate the backhaul communication congestion, reduce the computing delay, or a combination of both, depending on the networking scenario. Moreover, we develop a scalable optimization method to improve the algorithm efficiency for the proposed content placement scheme. Our method decouples and solves the complex content placement problem. Parallel content placement is enabled to select video chunks to cache from a large number of candidates with different formats and popularity correlations under resource constraints.

5.4.1   Adaptive Field-of-View Video Chunks

In this section, we introduce the formats of the video chunks for effective caching. As shown in Fig. 5.2, a full video is divided into VSs in the time domain. Then, a VS is further divided into smaller video chunks,in the unit of tiles, in the spatial domain. There are two existing solutions regarding the size for a VR video chunk:

- Tile-based caching solution [2, 7, 8]: Cloud and edge servers store the video chunks in the unit of tiles. Multiple video chunks are stitched together into an FoV-size video chunk and sent to user HMDs. The advantage is that the video quality in an FoV can be flexibly customized according to the network environment. Since users normally have a lower requirement on video quality for the edge of an FoV, the edge server can adjust the number of video chunks from the enhanced layer in the FoV, to adapt to the network environment. However, fine-granularity video chunks result in a low video encoding efficiency and a large amount of content to cache, which increases the overall storage usage required for caching and complicates caching management.
- FoV-based caching solution [9, 10, 22]: In this solution, the video chunk size is the same as the user's FoV size. Both 2D monoscopic and 3D stereoscopic video chunks can be cached at the edge server without stitching. However, the video quality for the tiles in a video chunk is fixed. Moreover, if FoVs with adjacent viewpoints are cached, the overlapping area of the FoVs in the cache causes redundancy and reduces the effectiveness of caching.

To balance caching efficiency and video quality adaptivity, we propose a hybrid content format solution, which combines the advantage of the two existing solutions.

5.4.1.1   Extended FoV

The video chunk setup is illustrated in Fig. 5.4. The spatial area of an FoV consists of $(\alpha_0 \times \beta_0)$ tiles. We encode and deliver video chunks with $(\alpha \times \beta)$ tiles to a user HMD, where $\alpha \geq \alpha_0$ and $\beta \geq \beta_0$. The corresponding 3D stereoscopic video chunk is referred to as an extended FoV (EFoV). The index of an EFoV is denoted by (i, s), where i follows the index of the tile located in the center of the video chunk, and s is the VS index for the video chunk. Denote the set of tiles associated with EFoVs $\{(i, s), \forall s\}$ by $\mathcal{T}_i$. Delivering a spatial area that is larger than an FoV can accommodate viewpoint movement variations and avoid frequent content delivery [6]. As a result, an EFoV can be used to render multiple adjacent viewpoints at a user HMD. The set of viewpoints that can be rendered by EFoV (i, s) is denoted by $\mathcal{V}_i$. In the example shown in Fig. 5.4, EFoV (i, s) can render all viewpoints located

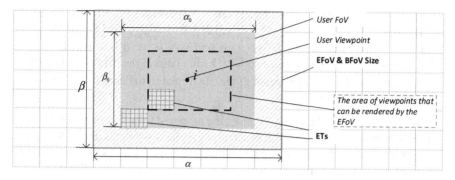

Fig. 5.4 The illustration of video chunk sizes and types

in the tiles around tile i, within the red dashed line. However, a large-size EFoV increases the data size of an EFoV. The parameters α and β depend on the video encoding efficiency and user viewpoint movement. Note that EFoVs are in 3D and ready to be downloaded by users.

5.4.1.2   Content Types

There are three types of video chunks that can be stored in the cache:

- Base FoV (BFoV): BFoVs are the 2D monoscopic video chunks, which cover the same spatial area as EFoVs.[1] The BFoV video chunks are partitions of the base layer of a video. Therefore, a BFoV has a low resolution and ensures the basic quality of service for users. The data size of BFoV (i, s) in VS s is denoted by $d_{i,s}^{BFoV}$ in the unit of byte;
- Enhanced tile (ET): An ET provides quality enhancement for a 2D monoscopic video chunk and covers the spatial area of a tile. ET video chunks are partitions of the enhanced layer of a video. ETs can be added on top of a BFoV to improve the video quality for a specific tile. When an FoV is requested by a user HMD, the BFoV and corresponding ETs are stitched together, projected to an EFoV, and sent to the user HMD. As the number of ETs added into the BFoV increases, the overall data size of projected EFoV increases correspondingly. The data size of ET (i, s) in VS s is denoted by $d_{i,s}^{ET}$;
- EFoV: As mentioned, EFoVs are 3D stereoscopic video chunks ready to be sent to user HMDs. EFoV (i, s) is projected from BFoV (i, s) combined with ETs (i', s), where $i' \in \mathcal{T}_i$. Thus, an EFoV video chunk may contain both base layer and enhanced layer video chunks. Video chunks projected from ET into

---

[1] The proposed scheme supports different BFoV image size through video processing, such as video stitching and clipping. Considering the video encoding efficiency and aiming to simplify the model, we assume that a BFoV here has the same image size as an EFoV.

3D format and added on an EFoV are referred to as 3D-ETs on the EFoV. The size of EFoV (i, s) is $d_{i,s}^{EFoV} = \omega_{i,s} d_{i,s}^{BFoV} + \sum_{i' \in T_i} \omega_{i',s}^{ET} \mathbf{I}_{i',i,s} d_{i,s}^{ET}$, where $\mathbf{I}_{i',i,s}$ indicates whether 3D-ET (i', s) is on EFoV (i, s). If it is true, $\mathbf{I}_{i',i,s} = 1$; otherwise, $\mathbf{I}_{i',i,s} = 0$. Let $\omega_{i,s}$ and $\omega_{i',s}^{ET}$ be the ratio of the data size after BFoV (i, s) and ET (i', s) are projected to the 3D format to the data size of the BFoV and the ET, respectively.

Similar to the tile-based caching solution, the proposed hybrid caching solution can adaptively change the number of ETs according to the network dynamics. Moreover, compared to the tile-based caching solution, we use video chunks that support base video quality in a large spatial area to improve the video encoding efficiency.

5.4.1.3 Rules for Content Distribution

As shown in Fig. 5.5, the three types of content, i.e., BFoV, ET, and EFoV, can be cached at the edge server. When a user HMD requests an EFoV to render viewpoint i for VS s, there are four cases:

- **Case 1: Direct delivery.** An EFoV that can render viewpoint i for VS s is cached. The video chunk is fetched from the cache and delivered to the user HMD directly;
- **Case 2: Projection and delivery.** No EFoV in the cache can render viewpoint i for VS s, while a BFoV that can render viewpoint i for VS s is cached. The BFoV is stitched with all the associated ETs cached in the edge server, projected to an EFoV, and delivered to the user HMD;
- **Case 3: Stitching, projection, and delivery.** No EFoV and BFoV in the cache can directly render viewpoint i for VS s. However, two or more BFoVs stored in the cache can be stitched together to render viewpoint i for VS s. The BFoVs are processed and projected to an EFoV. The EFoV is delivered to the user HMD;
- **Case 4: Fetching, stitching, projection, and delivery.** No EFoV and BFoV in the cache can render or be stitched to render the viewpoint. The corresponding BFoV is downloaded from the cloud server, stitched with any ET cached at the edge server, projected to an EFoV through a computing unit, and delivered to the user HMD.

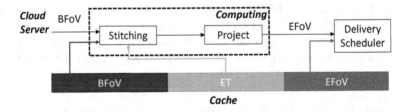

Fig. 5.5 Flowchart for content distribution at the edge server

5.4.2 *Content Placement on an Edge Cache*

In this subsection, we present a content placement scheme for caching popular and high-quality video chunks at the edge server. We first formulate the objective function for content placement. Taking into account the trade-off among multi-dimensional resources, we propose a content placement scheme for video chunks in a VS. Moreover, we extend the content placement scheme for placing the video chunks with different VSs.

The objective in content placement is to maximize the popularity of content in the cache, while

- Improving the video quality of the cached video chunks as much as possible;
- Satisfying an average content delivery latency requirement;
- Complying with the maximum cache capacity C.

Denote content placement variables by $e_{i,s}^{EFoV}$, $e_{i,i',s}^{EFoV-E}$, $e_{i,s}^{BFoV}$, and $e_{i,s}^{ET}$, $\forall i, s$, which indicate

- whether EFoV (i, s) is cached at the edge server;
- whether 3D-ET (i', s) corresponding to cached EFoV (i, s) is cached;
- whether BFoV (i, s) is cached at the edge server;
- whether ET (i, s) is cached at the edge server,

respectively. If the corresponding video chunk is cached, the corresponding indicator is one; otherwise, the indicator is zero. Let $\mathcal{E}_s = \{\mathcal{E}_s^{EFoV}, \mathcal{E}_s^{EFoV-E}, \mathcal{E}_s^{BFoV}, \mathcal{E}_s^{ET}\}$ denote the set of cached video chunks in VS s. To evaluate the quality of cached video chunks, we utilize a parameter, w, to weigh the popularity of video chunks from the enhanced layer. When $w > 1$, high video quality is required by a user; otherwise ($w \le 1$), the user has less strict demand on the video quality. We also introduce parameter $o_{i,i'}$ to represent the quality requirement for tile i' in EFoV (i, s). Denote the physical distance between tiles i and i' by $d_{i,i'}$. Since a lower quality is required for the marginal area of an FoV [6, 7], parameter $o_{i,i'}$ is proportional to $1/d_{i,i'}$. Let p_s^{EFoV} and p_s^{BFoV} denote the overall weighted popularity achieved by the cached EFoV and BFoV in VS s. We formulate the weighted popularity for the cached video chunks according to VS s as follows:

- The weighted popularity achieved by EFoV (i, s) can be formulated by

$$p_{i,s} = \sum_{i' \in \mathcal{V}_i^*} p_{i',s}^P (1 + \frac{w}{\alpha\beta} \sum_{m \in \mathcal{T}_i} o_{i,m} e_{i,m,s}^{EFoV-E}), \quad i \in \mathcal{E}_s^{EFoV} \tag{5.1}$$

where $\mathcal{V}_i^*$ is a subset of $\mathcal{V}_i$. It includes the viewpoints in set $\mathcal{V}_i$ which cannot be rendered by other cached EFoVs except EFoV (i, s). The weighted popularity of EFoV (i, s) is the summation of the popularity of viewpoints that can be rendered by caching the EFoV and the weighted popularity of 3D-ETs on the EFoV. For all cached EFoVs, the weighted popularity is

$$p_s^{EFoV} = \sum_{i'} p_{i',s}^P \max_{\{i|i'\in\mathcal{V}_i\}} e_{i,s}^{EFoV}(1 + \frac{w}{\alpha\beta} \sum_{m\in\mathcal{T}_i} o_{i',m} e_{i,m,s}^{EFoV-E}). \tag{5.2}$$

The overall weighted popularity contributed by caching EFoV files includes two parts: One is the overall popularity of viewpoints that can be rendered by the cached EFoVs; the other is the summation of the maximum weighted popularity contributed by 3D-ETs on the cached EFoVs to render viewpoints.

- The weighted popularity achieved by a BFoV (i, s) can be formulated by

$$p_{i,s} = \sum_{i'\in\mathcal{V}_i^*} p_{i',s}^P(1 + \frac{w}{\alpha\beta} \sum_{m\in\mathcal{T}_i} o_{i,m} e_{m,s}^{ET}), \quad i \in \mathcal{E}_s^{BFoV}. \tag{5.3}$$

Similar to the weighted popularity achieved by an EFoV, the weighted popularity achieved by a BFoV includes the summation of the popularity of viewpoints that can be rendered and the weighted popularity contributed by ETs. Note that ET (m, s) can be reused to add to any BFoV (i, s), where $i \in \{i|m \in \mathcal{T}_i\}$. BFoVs can be stitched together to render the viewpoints out of set $\mathcal{V}_i$ for BFoVs. Let $f_{i,s}^{BFoV}$ indicate if tile i is included by any BFoV stored in the cache, given by

$$f_{i,s}^{BFoV} = 1 - \prod_{\{m|m\in\mathcal{T}_i\}} (1 - e_{m,s}^{BFoV}). \tag{5.4}$$

A BFoV can be synthesized when all the tiles in the BFoV are included in the cached BFoVs. Moreover, since a BFoV can be projected and delivered only when a viewpoint cannot be rendered by any cached EFoV, only the popularity of viewpoints that cannot be rendered by EFoVs is taken into account. Therefore, for all cached BFoVs, the weighted popularity is

$$p_s^{BFoV} = \sum_{i'} p_{i',s}^P \prod_{\{i|i'\in\mathcal{V}_i\}} (1 - e_{i,s}^{EFoV})(\prod_{m\in\mathcal{T}_{i'}} f_{m,s}^{BFoV} + \frac{w}{\alpha\beta} \sum_{m\in\mathcal{T}_{i'}} o_{i,i'} e_{m,s}^{ET}). \tag{5.5}$$

Meanwhile, caching different types of video chunks results in different delivery delay. Given the channel Markov chain in Sect. 5.3.1, $\bar{R}_E$ can be obtained by

$$\bar{R}_E = \frac{p_H}{p_H + p_L} R_{E,H} + \frac{p_L}{p_H + p_L} R_{E,H}. \tag{5.6}$$

Let t_s^{EFoV}, t_s^{BFoV}, and t_s^{Cloud} denote the average content delivery delay for rendering viewpoints in VS s by the cached EFoVs, the cached BFoVs, and the video chunks from the cloud server, respectively. The average content delivery delay for rendering a viewpoint can be formulated as follows.

- For EFoV (i, s) cached at the edge server, the average content delivery delay is the video chunk transmission time from the edge server to a user HMD, given by

$$q_{i,s}^{EFoV} = (\omega_{i,s} d_{i,s}^{BFoV} + \sum_{i' \in \mathcal{T}_i} \omega_{i',s}^{ET} e_{i,i',s}^{EFoV-E} d_{i',s}^{ET}) \bar{R}_E^{-1}, \quad i \in \mathcal{E}_s^{EFoV}. \tag{5.7}$$

For all viewpoints in VS s that can be rendered by the cached EFoVs, the average delay is

$$t_s^{EFoV} = \sum_{i'} p_{i',s}^P \min_{\{i|i' \in \mathcal{V}_i\}} e_{i,s}^{EFoV} q_{i,s}^{EFoV}. \tag{5.8}$$

- For BFoV (i, s) cached at the edge server, the average content delivery delay includes the video processing time (for video stitching and projection) and the video chunk transmission time from the edge server to a user HMD, which is

$$q_{i,s}^{BFoV} = (\omega_{i,s} d_{i,s}^{BFoV} + \sum_{i' \in \mathcal{T}_i} \omega_{i',s}^{ET} e_{i',s}^{ET} d_{i',s}^{ET}) \bar{R}_E^{-1} + (d_{i,s}^{BFoV} + \sum_{i' \in \mathcal{T}_i} e_{i',s}^{ET} d_{i',s}^{ET}) R_C^{-1} \tag{5.9}$$

where R_C represents the computing rate for video processing. Denote the average number of computing cycles needed to process a data bit by χ. The average computing rate is $R_C = f/\chi$. For all viewpoints in VS s that can be rendered by cached BFoVs, the average delay is

$$t_s^{BFoV} = \sum_{i'} p_{i',s}^P [\prod_{\{i|i' \in \mathcal{V}_i\}} (1 - e_{i,s}^{EFoV})]\{ \min_{\{i|i' \in \mathcal{V}_i\}} q_{i,s}^{BFoV} e_{i,s}^{BFoV}$$
$$+ \prod_{\{i|i' \in \mathcal{V}_i\}} (1 - e_{i,s}^{BFoV})[(\prod_{i \in \mathcal{T}_{i'}} f_{i,s}^{BFoV}) q_{i',s}^{BFoV} e_{i',s}^{BFoV}]\}. \tag{5.10}$$

Equation (5.10) includes the average delay in Cases 2 and 3 of the content delivery scenario in Sect. 5.4.1.2: for Case 2, when viewpoint i' can be rendered by at least one of the cached BFoVs, the average delay is the minimum average delivery delay achieved by the cached BFoVs. For case 3, when viewpoint i' cannot be rendered by any of the cached BFoVs but can be stitched by multiple BFoVs, BFoV (i', s) will be stitched and delivered, and the average delivery delay is the delay for processing and transmitting BFoV (i', s).

- For viewpoints in VS s that cannot be rendered by any EFoV and BFoV in the cache, the corresponding BFoVs are downloaded from the cloud server. In such case, the average content delivery delay includes the time of downloading BFoV (i, s) from the cloud server to the edge server, the video processing time, and the video chunk transmission time from the edge server to the user HMD, given by

$$q_{i,s}^{Cloud} = q_{i,s}^{BFoV} + d_{i,s}^{BFoV} R_B^{-1}. \tag{5.11}$$

The average delay for delivering video chunks from the cloud server is

$$t_s^{Cloud} = \sum_{i'} p_{i',s}^P \prod_{\{i|i'\in\mathcal{V}_i\}} (1 - e_{i,s}^{EFoV})(1 - \prod_{i\in\mathcal{T}_{i'}} f_{i,s}^{BFoV})$$

$$\cdot (q_{i',s}^{BFoV} + d_{i',s}^{BFoV} R_B^{-1}). \tag{5.12}$$

In summary, the optimization problem of content placement can be formulated as

$$\max_{\{\mathcal{E}_s, \forall s\}} \sum_{s\in\mathcal{S}} p_s^{VS}(p_s^{EFoV} + p_s^{BFoV}) \tag{5.13a}$$

$$\text{s.t.} \sum_{s\in\mathcal{S}} p_s^{VS}(t_s^{EFoV} + t_s^{BFoV} + t_s^{Cloud}) \le \delta \tag{5.13b}$$

$$\sum_{s\in\mathcal{S}} \sum_{i\in\mathcal{I}} d_{i,s}^{BFoV}(\omega_{i,s} e_{i,s}^{EFoV} + e_{i,s}^{BFoV})$$

$$+ d_{i,s}^{ET}(\sum_{i'\in\mathcal{T}_i} \omega_{i',s}^{ET} e_{i,i',s}^{EFoV-E} + e_{i,s}^{ET}) \le C \tag{5.13c}$$

$$e_{i,i',s}^{EFoV-E} = 0, \text{ if } e_{i,s}^{EFoV} = 0 \tag{5.13d}$$

$$e_{i,s}^{EFoV}, e_{i,i',s}^{EFoV-E}, e_{i,s}^{BFoV}, e_{i,s}^{ET} \in \{0,1\}, \forall s, i, i' \tag{5.13e}$$

where δ is the average content delivery delay threshold. Solving the above optimization problem has two challenges. First, the problem is a non-linear combinatorial optimization problem. A similar problem is the maximal coverage problem [23], which is proved to be NP-hard. Second, the large number of video chunks results in a large number of variables. Consider a five-minute video for example. A one-second equirectangular video can be divided into 24×12 tiles/viewpoints in the spatial dimension. For each viewpoint, there are at least 4 decision variables for a VS, i.e., $e_{i,s}^{EFoV}$, $e_{i,s}^{BFoV}$, $e_{i,s}^{ET}$, and $\{e_{i,i',s}^{EFoV-E}, \forall i'\}$. Consequently, at least 345,600 decision variables need to be determined for caching this video, which leads to significantly high computing complexity. Therefore, we propose a two-stage solution for problem (5.13). We first decouple the problem into subproblems, which place content for each individual VS, given the delay bound and the cache capacity bound for VSs. We then determine the two bounds, i.e., δ_s and C_s, for all VSs by an iterative optimization method. Our proposed solution divides a large number of decision variables into a number of small variable sets, makes decisions for the small set of variables in a parallel manner, and attains the optimal solution iteratively.

5.4.3   Placement Scheme for Video Chunks in a VS

Video chunks in different VSs share cache capacity C, where the share of cache capacity for video chunks in VS s is C_s. We introduce a delay bound, δ_s, for caching video chunks in VS s. The values of C_s and δ_s are determined in Sect. 5.4.4. Given C_s and δ_s, problem (5.13) is decoupled into subproblems, each of which places video chunks in a VS. A subproblem is formulated as

$$\max_{\mathcal{E}_s} \; p_s^{EFoV} + p_s^{BFoV} \tag{5.14a}$$

$$\text{s.t.} \; t_s^{EFoV} + t_s^{BFoV} + t_s^{Cloud} \leq \delta_s \tag{5.14b}$$

$$\sum_{i \in \mathcal{I}} d_{i,s}^{BFoV} (\omega_{i,s} e_{i,s}^{EFoV} + e_{i,s}^{BFoV})$$

$$+ d_{i,s}^{ET} \left(\sum_{i' \in \mathcal{T}_i} \omega_{i',s}^{ET} e_{i,i',s}^{EFoV-E} + e_{i,s}^{ET} \right) \leq C_s \tag{5.14c}$$

$$e_{i,i',s}^{EFoV-E} = 0, \; \text{if } e_{i,s}^{EFoV} = 0 \tag{5.14d}$$

$$e_{i,s}^{EFoV}, e_{i,i',s}^{EFoV-E}, e_{i,s}^{BFoV}, e_{i,s}^{ET} \in \{0, 1\}, \forall i, i'. \tag{5.14e}$$

Denote the optimal weighted popularity for cached video chunks of VS s by $P_s(C_s, \delta_s)$. To solve the constrained problem (5.14), we propose a heuristic algorithm. The first step in the algorithm is to select video chunks to cache with the objective of guaranteeing delay constraints with minimum cache capacity usage. The second step is to cache popular content to maximize the objective function in (5.14a).

In the first step, we select video chunks that can effectively reduce the average content delivery delay. We define an index, h_i, to represent the average delay deduction per bit by caching video chunk i. The index of VS s is dropped since we focus on only one VS here. Note that 3D-ETs on EFoVs and ETs are not considered in this step since caching ETs cannot reduce the average content delivery delay. Based on (5.7), (5.9), and (5.11), h_i for EFoVs and BFoVs can be obtained as:

$$h_i = \begin{cases} p_{i,s}(R_C^{-1} + R_B^{-1})\omega_{i,s}^{-1}, & \text{for } i \in \mathcal{E}_s^{EFoV} \\ p_{i,s}(R_B^{-1}), & \text{for } i \in \mathcal{E}_s^{BFoV}. \end{cases} \tag{5.15}$$

A higher h_i means that caching video chunk i is more effective in reducing content delivery delay. The main idea of this step is to cache as many video chunks with high h_i as possible under delay bound δ_i. However, the problem is non-linear since multiple EFoVs can render the same viewpoint, and multiple BFoVs can be stitched to render a viewpoint. The value of $p_{i,s}$ depends on other video chunks in the cache, and so does index h_i. Let $H(\mathcal{E})$ represent the summation of index h_i achieved by video chunks in set $\mathcal{E}$. Without considering stitching BFoVs,

$H(\mathcal{E}_1 \cup \mathcal{E}_2) \leq H(\mathcal{E}_1) + H(\mathcal{E}_2) - H(\mathcal{E}_1 \cap \mathcal{E}_2)$, where strict inequality holds when a video chunk from $\mathcal{E}_1$ and a video chunk from $\mathcal{E}_2$ can render the same viewpoints. It makes the function $H(\mathcal{E})$ a sub-modular set function, and the problem is similar to a maximal coverage problem, in which a near-optimal solution can be obtained by a greedy approach [23, 24]. The corresponding greedy content placement scheme is presented in Algorithm 8, where the video chunk that maximizes $H(\mathcal{E})$ is selected consecutively as long as the average delay is not larger than the delay bound. Function $t_s(\mathcal{E})$ represents the content delivery delay after caching the content in $\mathcal{E}$, which can be obtained by (5.8), (5.10), and (5.12). Let $h(\{i\}|\mathcal{E})$ represent the index of video chunk i given that video chunks in set $\mathcal{E}$ are cached, where

$$h(\{i\}|\mathcal{E}) = H(\{i\} \cup \mathcal{E}) - H(\mathcal{E}). \tag{5.16}$$

In practice, adjacent BFoVs can be stitched to render a viewpoint. In such case, the sub-modularity of $H(\mathcal{E})$ no longer holds. Therefore, we design a heuristic approach to adjust the solution from Algorithm 8, which is provided in Algorithm 9, where k represents the iteration number, and $\mathcal{E}^{BFoV,k+1}$ represents the set of BFoVs in $\mathcal{E}^{k+1}$. The idea of the heuristic algorithm is to cache BFoVs to further improve the overall achieved index value, $H(\mathcal{E})$, until no BFoV can further improve $H(\mathcal{E})$. The algorithm provides a feasible content placement solution of problem (5.14). If the content placement result cannot satisfy the average delay requirement, we consider the content placement problem infeasible.

After a feasible solution is obtained, we further improve the overall weighted popularity of content in the cache, as presented in Algorithm 10. The idea is similar to that of Algorithm 9, while Algorithm 10 maximizes the overall weighted popularity rather than the overall index value. Let function $P(\mathcal{E})$ represent the overall weighted popularity by caching video chunks in set $\mathcal{E}$. The value of $P(\mathcal{E})$ can be obtained from objective function (5.14a). Let $p(\{i\}|\mathcal{E})$ represent the weighted

---

Algorithm 8 Greedy content placement

Input: Index function $H(\cdot)$, delay requirement δ_s, and cache capacity C_s.
Output: Set of video chunks $\mathcal{E}^0$.
1: $\mathcal{E}^0 \leftarrow \emptyset$.
2: $\hat{\mathcal{E}} \leftarrow$ all BFoV and EFoV video chunks.
3: **while** $\exists i \in \hat{\mathcal{E}}\}|\{t_s(\mathcal{E}^0 \cup \{i\}) \geq \delta_s\}$ **do**
4:    **if** the overall data size of content in $\mathcal{E}^0$ is greater than C_s **then**
5:       **Break**
6:    **end if**
7:    $i = \underset{i \in \hat{\mathcal{E}}}{\text{argmax }} h(\{i\}|\mathcal{E}^0)$.
8:    $\mathcal{E}^0 \leftarrow \mathcal{E}^0 \cup \{i\}$.
9:    $\hat{\mathcal{E}} \leftarrow \hat{\mathcal{E}} \backslash \{i\}$.
10: **end while**
11: Obtain the indexes of cached video chunks: $\{h(\{i\}|\mathcal{E}^0 \backslash \{i\}), \forall i \in \mathcal{E}^0\}$.
12: Return $\mathcal{E}^0$.

Algorithm 9 Heuristic content adjustment—Stage 1

Input: Index function $H(\cdot)$, delay bound δ_s, and cache capacity bound C_s.
Output: Feasible solution $\mathcal{E}$.

1: Run Algorithm 8, $k = 0$.
2: **while** 1 **do**
3:    Find a BFOV i^* not in $\mathcal{E}^k$, where:

$$i^* = \underset{i}{\mathrm{argmax}}\ h(\{i\}|\mathcal{E}^k\backslash\{i'\}), i' \in \mathcal{E}^{BFoV,k}, i \notin \mathcal{E}^{BFoV,k},$$

   and $h(\{i^*\}|\mathcal{E}^k\backslash\{i'\})$ is larger than the minimum index of video chunks in $\mathcal{E}^k$.
4:    **if** i^* exists **then**
5:        $\mathcal{E}^k \leftarrow \{\mathcal{E}^k\backslash\{i'\}\} \cup \{i^*\}, k = k + 1$.
6:    **else**
7:        **Break**. Return set $\mathcal{E}^k$.
8:    **end if**
9:    **while** the overall data size of content in $\mathcal{E}^k$ is greater than C_s **do**
10:        Delete the video chunk with the minimum index.
11:    **end while**
12: **end while**

Algorithm 10 Heuristic content adjustment—Stage 2

Input Index function $H(\cdot)$, delay bound δ_s, and cache bound C_s.
Output Content placement solution $\mathcal{E}_s$.

1: Run Algorithm 9 to obtain the feasible solution $\mathcal{E}$, and initialize $\mathcal{E}^0 = \mathcal{E}, k = 0$.
2: **while** 1 **do**
3:    Run Algorithm 8 to place content with the highest index $h\{\{i\}|\mathcal{E}^k\}$ until cache capacity bound C_s is reached.
4:    Find video chunk i^*, where:

$$i^* = \underset{i}{\mathrm{argmax}}\ p(\{i\}|\mathcal{E}^k\backslash\{i'\}), \forall i' \in \mathcal{E}^k, \forall i \notin \mathcal{E}^k$$

   and (a) $p(\{i^*\}|\mathcal{E}^k\backslash\{i'\})$ is larger than the minimum weighted popularity of video chunks in $\mathcal{E}^k$;
   (b) the average delay constraint is satisfied if video chunks i^* and i' are switched.
5:    **if** i^* exists **then**
6:        $\mathcal{E}^k \leftarrow \{\mathcal{E}^k\backslash\{i'\}\} \cup \{i^*\}, k = k + 1$.
7:    **else**
8:        **Break**. Return set $\mathcal{E}_s = \mathcal{E}^k$.
9:    **end if**
10:    Obtain the weighted popularity of video chunks in the cache: $\{p(\{i\}|\mathcal{E}^k\backslash\{i\}), \forall i \in \mathcal{E}^k\}$.
11:    **while** the overall data size of video chunks in $\mathcal{E}^k$ is larger than C_s **do**
12:        Delete video chunks with the minimum weighted popularity.
13:    **end while**
14: **end while**

---

popularity of video chunk i given that video chunks in set E are cached, where

$$p(\{i\}|\mathcal{E}) = P(\{i\} \cup \mathcal{E}) - P(\mathcal{E}). \tag{5.17}$$

The content placement is finished when there is no video chunk that can be cached to further improve the overall weighted popularity. In this case, a local optimum of problem (5.14) is achieved.

5.4.4   Placement Scheme for Video Chunks of Multiple VSs

Although Algorithm 10 can be used to cache video chunks of different VSs, high computational complexity is expected due to the large number of variables and the correlation in popularity among video chunks. Observing the independence among video chunks in different VSs, we next propose an optimization scheme to determine the share of cache capacity C and delay requirement δ for individual VSs, i.e., C_s and δ_s, respectively. Based on parameters C_s and δ_s, we can select the video chunks to cache from different VSs in a parallel manner.

Given C_s and δ_s, the solution of (5.14) for one VS can be obtained, i.e., $P_s(C_s, \delta_s)$. The overall optimization problem (5.13) can be rewritten as

$$\max_{\{C_s, \delta_s, \forall s\}} \sum_{s \in S} p_s^{VS} P_s(C_s, \delta_s) \tag{5.18a}$$

$$\text{s.t.} \sum_{s \in S} p_s^{VS} \delta_s \leq \delta \tag{5.18b}$$

$$\sum_{s \in S} C_s \leq C \tag{5.18c}$$

$$C_s \geq C_s^{min}(\delta_s), \forall s. \tag{5.18d}$$

In (5.18d), $C_s^{min}(\delta_s)$ represents the minimum cache resource required to cache the video chunks of VS s in order to satisfy the delay bound δ_s for delivering these video chunks. Function $P_s(C_s, \delta_s)$ depends on the viewpoint distribution and the data size of video chunks in VS s and is not known in advance. Although the value of $P_s(C_s, \delta_s)$ can be obtained by Algorithm 10 given C_s and δ_s, the continuous range of the two variables makes problem (5.18) intractable. Therefore, utilizing an approach similar to that in Sect. 5.4.3, we first determine a suboptimal solution for δ_s. After the value of δ_s for all VSs is known, we then determine $P_s(C_s, \delta_s)$ tentatively via an optimization method.

Both Algorithms 8 and 9 place the video chunk with the highest index $h(\{i\}|\mathcal{E}\backslash\{i\})$ as long as the average delay requirement is satisfied. Therefore, there exists a minimum index h^{min}. The average delay constraint can be satisfied by caching the video chunks with indexes higher than the minimum index. As the average delay bound decreases, the minimum index for caching increases. Similarly, for caching video chunks of multiple VSs, there exists a global minimum index to meet the delay requirement with δ. Thus, we find h^{min} to meet the delay bound. Since h^{min} does not increase with δ, we use the bisection method to find its minimum index. The index for the cached video chunk (i, s) is $h(\{\{(i, s)\}|\mathcal{E}\backslash\{(i, s)\}\})$. Using an idea similar to that in Algorithms 8 and 9, we place the content with indexes $h(\{\{(i, s)\}|\mathcal{E}\backslash\{(i, s)\}\})$, which are higher than h^{min}, in parallel for all VSs. The average delay bound, δ_s, and the corresponding required minimum storage resource, $C_s^{min}(\delta_s)$, can be obtained after placing the video chunks.

After obtaining δ_s and $C_s^{min}(\delta_s)$, we utilize an ADMM technique to find the cache capacity bound C_s for all VSs. The ADMM technique allows attaining the optimal solution iteratively and, more importantly, optimizing function $P_s(C_s, \delta_s)$ for all VSs in a parallel manner. Using ADMM, we can tentatively estimate the unknown function, $P_s(C_s, \delta_s)$, for each VS iteratively.

To decompose problem (5.18) for individual VSs, we introduce auxiliary variables z_1 and $\{z_{2,s}, \forall s\}$ and rewrite constraints (5.18c) and (5.18d) as

$$\bar{C} - \frac{C}{|\mathcal{S}|} - z_1 = 0 \tag{5.19a}$$

$$C_s - z_{2,s} = C_s^{min}(\delta_s), \forall s \tag{5.19b}$$

where $\bar{C}$ represents the mean value of $\{C_s, \forall s\}$. Correspondingly, the augmented Lagrangian of problem (5.18) is formulated as

$$\mathcal{L}(\{C_s, z_{2,s}, u_{2,s}, \forall s\}, z_1, u_1) = -\sum_{s \in \mathcal{S}} p_s^{VS} P_s(C_s, \delta_s) + u_1(\bar{C} - \frac{C}{|\mathcal{S}|} - z_1)$$

$$+ \sum_s u_{2,s}(C_s - z_{2,s}) + \frac{\rho_1}{2}(\bar{C} - \frac{C}{|\mathcal{S}|} - z_1)^2$$

$$+ \frac{\rho_2}{2} \sum_s (C_s - z_{2,s})^2 \tag{5.20}$$

where u_1 and $\{u_{2,s}, \forall s\}$ are the Lagrange multipliers for constraints (5.19a) and (5.19b), respectively. Parameters ρ_1 and ρ_2 are penalty factors for augmented Lagrangian. It is straightforward that $P_s(C_s, \delta_s)$ is a non-decreasing function of C_s (i.e., as the cache size increases, the overall weighted popularity for video chunks in the cache increases). Therefore, $P_s(C_s, \delta_s)$ is a quasi-convex function. To prevent the algorithm from converging to a local optimum, a variation of ADMM, i.e., relaxed heavy ball ADMM [25], is adopted. Denote the iteration number by k. The variables in (5.20) can be updated iteratively as follows:

$$C_s^{k+1} = \mathrm{argmin}_{C_s^{k+1}}\{-p_s^{VS} P_s(C_s^{k+1}, \delta_s) + u_1^k(\bar{C}^k - C_s^k + C_s^{k+1} - \frac{C}{|\mathcal{S}|} - z_1^k)$$

$$+ u_{2,s}^k(C_s^{k+1} - z_{2,s}) + \frac{\rho_1}{2}(\bar{C}^k - C_s^k + C_s^{k+1} - \frac{C}{|\mathcal{S}|} - z_1^k)^2$$

$$+ \frac{\rho_2}{2}(C_s^{k+1} - z_{2,s}^k)^2\} \tag{5.21a}$$

$$z_1^{k+1} = \min\{\varepsilon(\bar{C}^{k+1} - \frac{C}{|\mathcal{S}|}) + (1-\varepsilon)\hat{z}_1^k + \frac{1}{\rho_1}\hat{u}_1^k, 0\} \tag{5.21b}$$

$$z_{2,s}^{k+1} = \max\{\varepsilon C_s^{k+1} + (1-\varepsilon)\hat{z}_{2,s}^k + \frac{1}{\rho_2}\hat{u}_{2,s}^k, C_s^{min}(\delta_s)\}, \forall s \tag{5.21c}$$

$$u_1^{k+1} = \hat{u}_1^k + \rho_1[\varepsilon(\bar{C}^{k+1} - \frac{C}{|\mathcal{S}|}) + (1-\varepsilon)\hat{z}_1^k - z_1^{k+1}] \tag{5.21d}$$

$$u_{2,s}^{k+1} = \hat{u}_{2,s}^k + \rho_1[\varepsilon C_s^{k+1} + (1-\varepsilon)\hat{z}_{2,s}^k - z_{2,s}^{k+1}], \forall s \tag{5.21e}$$

$$\hat{u}_1^{k+1} = u_1^{k+1} + \gamma(u_1^{k+1} - \hat{u}_1^k) \tag{5.21f}$$

$$\hat{u}_{2,s}^{k+1} = u_{2,s}^{k+1} + \gamma(u_{2,s}^{k+1} - \hat{u}_{2,s}^k), \forall s \tag{5.21g}$$

$$\hat{z}_1^{k+1} = z_1^{k+1} + \gamma(z_1^{k+1} - \hat{z}_1^k) \tag{5.21h}$$

$$\hat{z}_{2,s}^{k+1} = z_{2,s}^{k+1} + \gamma(z_{2,s}^{k+1} - \hat{z}_{2,s}^k), \forall s \tag{5.21i}$$

where $0 \le \varepsilon \le 1$, and $0 \le \gamma \le 1$. Solving (5.20) requires the knowledge of the gradient of $P_s(C_s, \delta_s)$. We introduce a non-decreasing function, $Q_s(C_s)$, to estimate $P_s(C_s, \delta_s)$ iteratively, where δ_s is omitted for simplicity. Equation (5.21a) is rewritten as

$$C_s^{k+1} = \operatorname{argmin}_{C_s^{k+1}}\{-p_s^{VS}Q_s(C_s^{k+1}) + u_1^k(\bar{C}^k - C_s^k + C_s^{k+1} - \frac{C}{|\mathcal{S}|} - z_1^k)$$

$$+ u_{2,s}^k(C_s^{k+1} - z_{2,s}) + \frac{\rho_1}{2}(\bar{C}^k - C_s^k + C_s^{k+1} - \frac{C}{|\mathcal{S}|} - z_1^k)^2$$

$$+ \frac{\rho_2}{2}(C_s^{k+1} - z_{2,s}^k)^2\} \tag{5.22}$$

Consider a piecewise linear function for $Q_s(C_s)$. In iteration k, when C_s^{k+1} is obtained by (5.22), we place the content according to Algorithm 10 for the VSs in parallel and observe the true value of weighted popularity $P_s(C_s^{k+1}, \delta_s)$. Accordingly, $Q_s(C_s)$ is updated by

$$Q_s^{k+1}(C_s) = \begin{cases} \frac{Q_s^{k+1}(C_s^{k+1}) - Q_s^k(C_s^-)}{C_s^{k+1} - C_s^-}(C_s - C_s^-) + Q_s^k(C_s^-), \\ \qquad\qquad\qquad\qquad\qquad \text{for } C_s^- \le C_s < C_s^{k+1} \\ \frac{Q_s^k(C_s^+) - Q_s^{k+1}(C_s^{k+1})}{C_s^+ - C_s^{k+1}}(C_s^+ - C_s) + Q_s^{k+1}(C_s^{k+1}), \\ \qquad\qquad\qquad\qquad\qquad \text{for } C_s^{k+1} \le C_s < C_s^+ \\ Q_s^k(C_s), \qquad\qquad\qquad\qquad \text{Otherwise} \end{cases}$$

$$\tag{5.23}$$

where C_s^- and C_s^+ represent the minimum closest and maximum closest values obtained in the previous iterations compared to C_s^{k+1}, respectively. As the iteration number increases, $Q_s(C_s)$ approaches $P_s(C_s, \delta_s)$. For limiting influence from inaccurate estimation in the optimization process, we apply an attenuated noise on C_s^{k+1} to explore the real value of the original function. Let $\mathcal{N}(0, \sigma^2)$ represent a Gaussian random noise with zero mean and standard deviation σ. The rate of noise attenuation is α_{attn}. The algorithm for placing video chunks of multiple VSs is given in Algorithm 11.

The complexity of the proposed content placement scheme is analyzed as follows: For placing video chunks in one VS (Algorithm 10), the worst time complexity is $(M_1 \times N_g^2)$, where N_g is the number of video chunks in a VS and M_1 is the iteration number for searching the optimal result. For placing video chunks in multiple VSs (Algorithm 11), the worst time complexity is $(M_1 \times |\mathcal{S}| \times M_2 \times N_g^2)$, where M_2 is the number of iterations for Algorithm 11 to converge. Despite the iterations in Algorithm 11, the content placement for different VSs is in parallel. Moreover, if the weighted popularity functions $\{P_s(C_s, \delta_s), \forall s \in \mathcal{S}\}$ are known, the complexity of the ADMM algorithm is negligible, and the time complexity of Algorithm 11 for a parallel placement process becomes $(M_2 \times N_g^2)$, i.e., the time complexity of Algorithm 10.

Algorithm 11 Content placement scheme for video chunks of multiple VSs

---

    % Bisection method to determine δ_s
1: **while** 1 **do**
2:    $h = (h_{max} - h_{min})/2$.
3:    **for** VS s in $\mathcal{S}$ **do**
4:        Place and adjust video chunks with indexes $h((i, s)|\mathcal{E}\backslash\{(i, s)\}) \geq h$ by Algorithms 8 and 9.
5:    **end for**
6:    Calculate the overall average content delivery delay $t = \sum_{s \in \mathcal{S}} P_s^{VS}(t_s^{EFoV} + t_s^{BFoV} + t_s^{Cloud})$.
7:    If $|t - \delta|$ is smaller than a threshold: **Break**.
8:    If $t \leq \delta$: $h_{min} = h$. Otherwise: $h_{max} = h$.
9: **end while**
10: Calculate the content delivery delay for VS s, i.e., δ_s, and the content data size, i.e., $C_s^{min}(\delta_s)$, for all $s \in \mathcal{S}$.
    % ADMM method to determine C_s
11: Initialize estimation function $Q_s^0(C_s)$, $k = 0$.
12: **while** 1 **do**
13:    Calculate $C_s^{k+1}\{Q_s^k(C_s)\}$ by (5.22).
14:    Update axillary variables and Lagrange multipliers by (5.21b)–(5.21i).
15:    $C_s^{k+1} = C_s^{k+1} + \mathcal{N}(0, \sigma^2)$, $\forall s$.
16:    **for** VS s in $\mathcal{S}$ **do**
17:        Place video chunks of VS s given bounds C_s^{k+1} and δ_s by Algorithm 10.
18:        Let $Q_s^{k+1}(C_s^{k+1}) = P(C_s^{k+1}, \delta_s)$.
19:    **end for**
20:    Update $Q_s^{k+1}(C_s)$ by (5.23).
21:    $\sigma = \alpha_{attn}\sigma$, $k = k + 1$.
22:    **if** $|\mathcal{L}(\{C_s, z_{2,s}, u_{2,s}, \forall s\}, z_1, u_1)^{k+1} - \mathcal{L}(\{C_s, z_{2,s}, u_{2,s}, \forall s\}, z_1, u_1)^k|$ and σ are less than the thresholds **then**
23:        **Break**. Return the content placement solution with cache capacity bound C_s.
24:    **end if**
25: **end while**

Table 5.1 Video metadata
[11]

No	Video length	Content	Category
1	2'44"	Conan360-Sandwich	Performance
2	3'20"	Freestyle Skiing	Sport
3	4'48"	Google Spotlight-Help	Film

5.4.5   Numerical Results

In this section, we demonstrate the numerical results of the proposed content placement schemes.

Parameter Settings In the simulation, we utilize the HMD tracking dataset composed of 48 users watching 3 spherical VR videos with different categories [11]. The details of the VR videos are provided in Table 5.1. In the data set, user viewpoints in the 3D coordinate system are sampled once every millisecond. We map the viewpoint locations into 2D space via equirectangular projection to obtain the viewpoint popularity distribution in each tile. The $360° \times 180°$ equirectangular video is divided into 24×12 tiles. The user's FoV spans 7×5 tiles, and the BFoV and EFoV both span 9×7 tiles. The time length of a video chunk is 4 s. Therefore, the three videos have 41, 50, and 72 VSs, respectively. We randomly generate the data size of video chunks using Gaussian distributions, where the average data size of a BFoV and an ET are 1 and 0.1 MByte, respectively, and the standard deviation of the data size are 0.45 and 0.14 MByte, respectively. The ratios, $\omega_{i,s}$ and $\omega_{i,s}^{ET}$, are generated using a Gaussian distribution, both with mean and variance of 1.5 and 0.3, respectively. The communication rate R_B between the cloud and the edge server is 200 MB/s. The average content delivery delay bound, δ, is 85 ms.

Content Placement Performance We present the numerical results of the parallel content placement scheme for a number of VSs. The content placement scheme is presented in Algorithm 11. The first 40 VSs in videos 1 and 2 are evaluated in the simulation. The popularity of the 40 VSs in the two videos, i.e., p_s^{VS}, is generated randomly. The overall cache capacity is 380 MBytes, and R_C and $\bar{R}_E$ are 35 MBytes and 200 Mb, respectively. The required average delay bound is 85 ms.

The relation between the VS popularity, i.e., p_s^{VS}, and the cache capacity allocated for a VS, i.e., C_s, in videos 1 and 2 is shown in Figs. 5.6 and 5.7, respectively. The circle markers show the VS popularity sorted in an increasing order. The cross markers show the corresponding allocated cache capacity, and the markers are fitted to a second-order polynomial line. As the popularity of a VS increases, the cache capacity assigned to the VS increases correspondingly, as expected. However, due to different viewpoint distributions in different VSs, the allocated cache capacities among VSs are different. Moreover, compared to video 1, the correlation between the popularity and the allocated cache capacity for a VS is weak, especially for the VSs with high popularity. The reason is, compared to video 1 (performance video), video 2 (film video) has a more concentrated viewpoint distribution. The optimal overall weighted popularity and the average

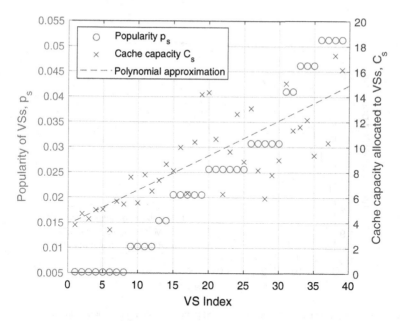

Fig. 5.6 The VS popularity p_s^{VS} and cache capacity allocated for a VS C_s for the first 40 VSs in video 1

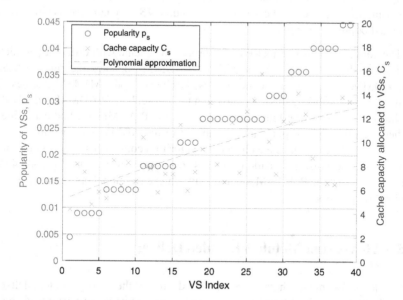

Fig. 5.7 The VS popularity p_s^{VS} and cache capacity allocated for a VS C_s for the first 40 VSs in video 2

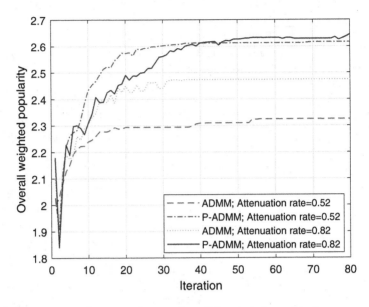

Fig. 5.8 Convergence performance of proposed parallel content placement scheme

delay constraints can be achieved by caching fewer video chunks. In this case, increasing the allocated cache size for the popular VS may not increase the overall weighted popularity significantly.

The convergence performance of the proposed scheme with different parameter settings is shown in Fig. 5.8, in which "P-ADMM" denotes the relaxed heavy ball ADMM method used in the proposed scheme, and "ADMM" denotes the conventional ADMM method. As shown in the figure, both P-ADMM and ADMM methods converge after 60 iterations. However, the P-ADMM method can achieve higher overall weighted popularity compared to the ADMM method since the P-ADMM method can reduce the chance the algorithm converges to a local optimum. Meanwhile, the noise attenuation rate has less impact in the case of P-ADMM, which shows that the P-ADMM method may achieve the global optimum via a short exploration process.

5.5   AI-Assisted Mobile VR Video Delivery

The content placement scheme in Sect. 5.4.2 decreases the average content delivery delay. However, the real-time video playback missing rate is not guaranteed so far, especially when unpopular video chunks are requested. In this section, we propose a content distribution framework to minimize video playback frame missing rate by distributing video chunks proactively. On the user side, the video chunks to be played in the future are requested according to the user viewpoint trajectory. On

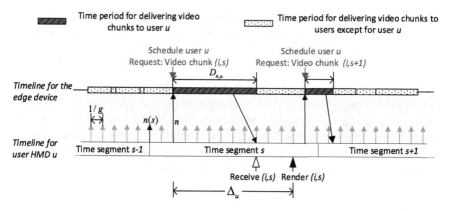

Fig. 5.9 Requesting and distributing video chunks for user u

the edge side, we propose a Whittle index (WI) based content delivery scheduling method to schedule delivery sequence according to user viewpoint movement, channel quality, and delivery time of requested video chunks.

5.5.1 Content Distribution

The content distribution framework is illustrated in Fig. 5.9. As mentioned in Sect. 5.3, the user viewpoint is sampled with frequency g. The n-th sampling interval is referred to as time slot n. The video watching timeline is divided into time segments that are aligned with the playback time of VSs. Thus, a time segment includes $\lceil T_s g \rceil$ slots. The index of the first time slot of time segment s is denoted by $n(s)$. In each time slot, the steps for requesting and delivering content in each time slots are summarized as follows:

(1) Each user HMD samples the user's viewpoint at the beginning of the time slot and checks if the viewpoint can be rendered by any downloaded EFoVs in the HMD's buffer;

(2) If the current viewpoint can be rendered, the user HMD plays the corresponding EFoV; otherwise, it plays the downloaded EFoV with a center point closest to the current viewpoint;

(3) The user HMD predicts the viewpoint in the subsequent time slots until finding a viewpoint that cannot be rendered by any EFoVs in the HMD's buffer;

(4) The user HMD sends a request to the edge server for an EFoV to render that viewpoint;

(5) The edge server, after receiving requests from all user HMDs, schedules a content delivery request by assigning a computing unit for processing the request. If no computing unit is available, the edge server cannot satisfy any request until at least a computing unit is available;

(6) If the request of a user HMD is scheduled, the edge server fetches the video
    chunk from the cache or the cloud, processes the video chunk, and sends
    the video chunk to the user HMD. The content delivery process occupies a
    computing unit. The user HMD saves the received video chunk in its buffer.

Note that if a request is not scheduled in a slot, the user HMD may request the
same EFoV in the subsequent slots until either of the following two cases happen.
First, the time slot has passed for the required EFoV, and another EFoV will be
required for a future time slot. Second, the user HMD identifies the prediction was
wrong, and the EFoV is no longer required for subsequent time slots. The detailed
content distribution settings are provided in the following subsection.

5.5.2   Intelligent Content Distribution Framework

The user HMDs predict the viewpoint movement of the users and request a video
chunk that they will watch in the subsequent time slots. If a request is scheduled
by the edge server, the requested video chunk is processed and delivered to the user
HMD. The received video chunk is then buffered in the device. In each slot, a user
HMD plays the video chunk corresponding to the user viewpoint if the video chunk
is stored at its buffer. If the viewpoint cannot be rendered by any video chunk in the
buffer, frame missing happens for that slot. The viewpoint of user u in time slot n is
denoted by $j_{u,n}$.

The set of video chunks in the buffer of user u in time slot n is denoted by $\mathcal{B}_{u,n}$.
Let function $\mathbb{A}(\mathcal{B}_{u,n})$ represent the set of viewpoints that can be rendered by the
video chunks in set $\mathcal{B}_{u,n}$. Thus, if viewpoint $j_{u,n}$ can be rendered by one video chunk
in the buffer, we have $j_{u,n} \in \mathbb{A}(\mathcal{B}_{u,n})$. Note that a video chunk has playtime T_s and
can render viewpoints in multiple time slots. The video chunks for the previous time
segments will be deleted automatically in the buffer.

Viewpoint prediction techniques, such as linear regression [7] and neural networks [13], have been widely investigated in the literature. In our scheme, we use
a LSTM neural network to predict future viewpoints according to the previous
user viewpoint trajectory. Nevertheless, any other AI-based viewpoint prediction
solution can be adopted in our proposed scheduling scheme. The previous viewpoint
trajectory of user u is denoted by $\mathbf{v}_{u,n} = \{j_{u,n-W}, \ldots, j_{u,n}\}$, where W denotes the
viewpoint window length for prediction. Given the trajectory, the user HMD predicts
the viewpoint in the next slot and checks if it can be rendered by any video chunks
stored in the buffer. If so, the user HMD consecutively predicts viewpoints in future
time slots until finding a viewpoint that cannot be rendered by any video chunks in
the buffer. The user HMD then send a request for the corresponding video chunks to
the edge server. The time between the current slot and the slot when the user HMD
will play the requested video chunk is denoted by Δ_u, which is referred to as request
duration.

We assume that user HMD has knowledge of the cached content at the edge server. Since multiple video chunks can render the same viewpoint, user HMDs select the video chunks with the shortest average content delivery time in their requests. For example, if both chunk A, cached as an EFoV, and B, cached as a BFoV, can render the future viewpoint, the user HMD requests chunk A. The video chunk that is selected and requested by user HMD u in time slot n for VS s is denoted by $(i, s)_{u,n}$.

For simplicity, we consider only one computing unit in the edge server, and the edge server can only process and deliver one request at a time. However, the proposed framework and solution can be extended to schedule delivery with multiple computing units. The edge server receives requests from user HMDs, selects one request, and delivers the requested video chunks to the corresponding user HMD. When a video chunk is scheduled for delivery, the edge server stitches all cached and associated ETs of the video chunk. Whenever the edge server finishes the delivery for one request, it evaluates received requests and selects one request for the next delivery. Different video chunk delivery requests result in different content delivery time. The content delivery time depends on the wireless channel condition and the content caching policy. We use the content delivery delay formulated in Sect. 5.4.2 to model the delivery time. The wireless transmission rate follows the two-stage Markov chain as mentioned in Sect. 5.3. The transmission rate in the n-th time slot for user HMD u is denoted by $R_{E,n,u}$. For notation simplicity, the overall number of time slots for delivering video chunks of the request scheduled in slot n for user HMD u is denoted by $D_{n,u}$.

The probability that a video chunk in the buffer matches the current viewpoint is referred to as hit probability. The objective in the content distribution phase is to schedule the content delivery sequence to minimize the frame missing likelihood, which can be achieved by maximizing the hit probability. Define scheduling variable $z_{u,n}$, where $z_{u,n} = 1$ if the request from user HMD u in time slot n is scheduled, and $z_{u,n} = 0$ otherwise. An optimization problem can be formulated as

$$\max_{\{z_{u,n}, \forall u,n\}} \lim_{N\to\infty} \frac{1}{N}\mathbb{E}\Big[\sum_n \sum_u 1\{j_{u,n} \in \mathbb{A}(\mathcal{B}_{u,n})\}\Big]. \qquad (5.24)$$

In (5.24), $1\{x\} = 1$ when x is true; $1\{x\} = 0$, otherwise. Meanwhile, the scheduling decision is constrained since only one request can be processed at any time. In order to solve the problem in a tractable manner, the constraint is relaxed into a time-averaged constraint given by

$$\lim_{N\to\infty} \frac{1}{N}\mathbb{E}\Big[\sum_n \sum_u z_{u,n}\Big] \le \frac{1}{\delta \times g}. \qquad (5.25)$$

The relaxed constraint indicates that at most one request can be scheduled during every $\delta \times g$ time slots, where $\delta \times g$ is the average number of slots for delivering a video chunk.

The scheduling decision depends on several factors. Intuitively, the video chunk in the request with the minimum Δ_u value is most urgently needed. However, if the user experiences poor channel quality or the video chunk has a long content delivery time, the requested video chunk may not be delivered in time. Furthermore, when the user viewpoint moves rapidly, e.g., when watching a sports game, the requested video chunk may render viewpoints only for a few slots. In contrast, when the user viewpoint moves slowly, the requested video chunk can support the viewpoints for multiple slots. When the computing resource is limited and only one request can be satisfied, the edge server may satisfy the request from the user HMD with lower viewpoint movement variation to increase the overall hit probability. Further, frequently changing viewpoints introduces a challenge for prediction accuracy. The uncertainty increases the possibility of a delivery failure, which should be considered in scheduling.

In the following two subsections, we reformulate the scheduling problem as a restless multi-armed bandit (RMAB) problem and propose a WI-based content delivery scheduling scheme. We further utilize a reinforcement learning method to determine the WI values in the scheduling scheme.

5.5.3 WI-based Delivery Scheduling

We first formulate the scheduling problem as an RMAB problem. Consider U controlled Markov chains $\{X_n^u, u = 1, \ldots, U, n \geq 0\}$ with state space $\mathcal{Y}^u$. The state of the Markov chain for user HMD u, denoted by Y_n^u, includes the previous viewpoint trajectory of user u, $\mathbf{v}_{u,n}$, the set of video chunks in the buffer of user HMD u, $\mathbf{B}_{u,n}$, and current channel condition, $R_{E,n,u}$. We add the information of the request in the state, including the requested video chunk index, $(i, s)_{u,n}$, the request duration, $\Delta_{u,n}$, the data size of the corresponding video chunk, and the content placement policy of the video chunk. The control variable for the Markov chain for user HMD u is $z_{u,n}$, which is binary: active when $z_{u,n} = 1$ and passive when $z_{u,n} = 0$. Denote a processing epoch by ϕ, where $\phi = D(u^*, n)$, in which u^* is the user HMD whose request is scheduled in time slot n. The probability of transiting from state Y_n^u to state $Y_{n+\phi}^u$ under control $z_{u,n}$ is denoted by $P(Y_{n+\phi}^u | Y_n^u, z_{u,n})$, where $\sum_{Y_{n+\phi}^u \in \mathcal{Y}^u} P(Y_{n+\phi}^u | Y_n^u, z_{u,n}) = 1$.

The reward for scheduling a request is the number of future viewpoints that can be rendered from granting that request, which is formulated as

$$R(Y_n^u, z_{u,n}) = \begin{cases} \sum_{n'=\min\{n+D_{n,u}, n(s)\}}^{n(s+1)-1} \kappa^{n'-n} \mathbf{1}\{j_{u,n'} \in \mathbb{A}(i_{u,n,s})\}, & \text{if } z_{u,n} = 1 \\ 0, & \text{if } z_{u,n} = 0. \end{cases}$$

$$(5.26)$$

The discounted reward model is applied in (5.26), where κ is the discount factor. When the discount factor is one, the average of the accumulated reward in an infinite time horizon for all user HMDs is equivalent to the average hit probability in (5.24). The problem is now an RMAB problem. Similar to the multi-armed bandit problem, in each decision epoch, we select one user HMD out of U user HMDs to schedule with unknown dynamics. Compared to multi-armed bandit problems, RMAB allows state transition when an arm is passive. In our case, the state of a user HMD still evolves even if the user HMD is not scheduled.

We solve the RMAB problem by a WI-based method, which is a heuristic solution and has good empirical performance [26]. Specifically, the WI-based method defines a *subsidy for passivity* for candidates, i.e., user HMDs, in the scheduling problem. The subsidy depends on the current state of each user HMD. The edge server will receive a higher long-term reward to schedule the user HMD with a higher subsidy. By the WI-based method, user HMDs can evaluate and transform the network dynamics into a WI value in a distributed manner. The edge server makes adaptive scheduling decisions by comparing the WI values collected from user HMDs. Denote the WI for user HMD u with state Y_n^u by $\lambda^*(Y_n^u)$. The WI can be obtained by

$$
\lambda^*(Y_n^u) = R(Y_n^u, 1) + \sum_{Y_{n+\phi}^u \in \mathcal{Y}^u} P(Y_{n+\phi}^u | Y_n^u, 1) \kappa^{D(n,u)} \max_{z_{u,n+\phi}} \{Q(Y_{n+\phi}^u, z_{u,n+\phi})\}
$$

$$
- \sum_{Y_{n+\phi}^u \in \mathcal{Y}^u} P(Y_{n+\phi}^u | Y_n^u, 0) \kappa^{\delta g} \max_{z_{u,n+\phi}} \{Q(Y_{n+\phi}^u, z_{u,n+\phi})\}. \tag{5.27}
$$

In (5.27), $Q(Y_n^u, z_{u,n})$ is the state-action value, i.e., Q value, for state Y_n^u with action $z_{u,n}$. The edge server collects the WI from all user HMDs and schedules the request with the maximum WI value. The conventional solution for finding WI cannot be applied in our considered problem since the state-action transition probability is unknown. The work [27] adopts the Q-learning method to learn the Q value and approximate the WI value. However, it cannot be applied to our problem due to the large and continuous state space. Therefore, we propose a novel deep reinforcement learning method to approximate the WI value.

5.5.4 Reinforcement Learning Assisted Content Distribution

Motivated by the actor-critic method, we adopt two neural networks. One of the neural networks approximates the Q value. Meanwhile, different from the actor-critic method, another neural network evaluates the WI value rather than provides the scheduling decision. Each user HMD can evaluate the Q value and WI in a distributed manner without information of other user HMDs.

Denote the weight vectors of the neural networks for evaluating the Q value and the WI value by θ_Q and θ_W, respectively. In time slot n, if the request of a user HMD is scheduled, the user HMD observes the reward as given in (5.26). Otherwise, the user HMD uses subsidy $\lambda^*(Y_n^u; \theta_W)$ as the reward, which represents the value of WI approximated by the neural network with weight vector θ_W. The weight vector θ_Q is updated by minimizing the loss function for evaluating the Q value, given by

$$L(\theta_Q) = \begin{cases} [Q(Y_n^u, 0; \theta_Q) - \lambda^*(Y_n^u; \theta_W) - Q(Y_{n+\phi}^u, 0; \theta_Q)]^2, & \text{if } z_{u,n} = 0 \\ [Q(Y_n^u, 1; \theta_Q) - R(Y_n^u, 1) - Q(Y_{n+\phi}^u, 1; \theta_Q)]^2, & \text{if } z_{u,n} = 1. \end{cases}$$
$$(5.28)$$

In (5.28), $Q(Y_{n+\phi}^u, z_{u,n}; \theta_Q)$ denotes the Q value approximated by the neural network with weight vector θ_Q. Based on the Q value, the WI value can be approximated in an iterative manner, given by

$$\hat{\lambda}^*(Y_n^u) = \lambda^*(Y_n^u; \theta_W) - \varphi_\lambda [Q(Y_n^u, 0; \theta_Q) - Q(Y_n^u, 1; \theta_Q)] \qquad (5.29)$$

where φ_λ denotes the step size for approximating the WI value. The true value of $\lambda^*(Y_n^u)$ satisfies $Q(Y_n^u, 0; \theta_Q) = Q(Y_n^u, 1; \theta_Q)$. Therefore, for the neural network with weight vector θ_W, the loss function for evaluating the WI value is

$$L(\theta_W) = [\lambda^*(Y_n^u; \theta_W) - \hat{\lambda}^*(Y_n^u)]^2. \qquad (5.30)$$

The proposed the WI-based content delivery is summarized in Algorithm 12, where φ_Q and φ_W are the learning rates for evaluating weight vectors θ_Q and θ_W, respectively. The algorithm has two parts: In the first part, presented in Lines 4 to 10, each user HMD evaluates its WI value by the neural networks in a distributed manner, and the device schedules the requests by selecting the request with the highest WI value. In the second part, presented in Lines 11 to 15, each user HMD trains the two neural networks based on the received reward. The weight vectors φ_Q and φ_W are updated in a distributed manner by minimizing loss functions (5.28) and (5.30). To ensure stability on learning, φ_Q should be higher than φ_W.

5.5.5   Neural Network Structure

The overview of the proposed learning-based content distribution scheme is shown in Fig. 5.10. In each time slot, user HMDs send requests for video chunks and the corresponding WI values. The edge server schedules the one with the highest WI value and delivers the corresponding video chunk. In each time slot, a user HMD observes its current state, predicts the future viewpoints by a future viewpoint prediction function, and requests the corresponding video chunk. Specifically, we apply an LSTM neural network to predict user viewpoints for the subsequent slots, which is illustrated on the left-hand side in Fig. 5.11. Moreover, the full state is evaluated by the Q value and WI evaluation networks. The WI value of the state will be sent to the edge server for scheduling content delivery, and the Q value of

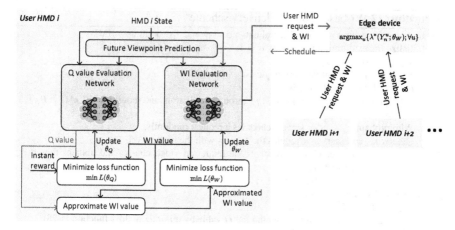

Fig. 5.10 The overview of the proposed learning-based content distribution scheme

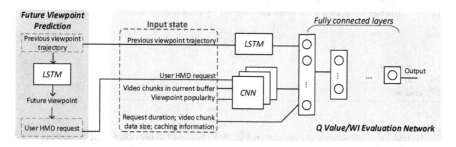

Fig. 5.11 Illustration of neural networks

the state will be utilized for estimating the loss functions and training the neural networks.

The neural network structures for the Q value and WI evaluation networks are similar, which are illustrated on the right-hand side in Fig. 5.11. We adopt an LSTM layer for finding the temporal correlation from the previous viewpoint trajectory and reducing the input data dimension. Moreover, both the request, $(i, s)_{u,n}$, and the set of video chunks in the buffer, $\mathcal{B}_{u,n}$, are represented by matrices with size $X \times Y$, and each element corresponds to a tile in the equirectangular size video. In the matrices representing $(i, s)_{u,n}$ and $\mathcal{B}_{u,n}$, an element is 1 if the location of the element is the same with the location of tile i and the location of a viewpoint that can be rendered by EFoVs in set $\mathcal{B}_{u,n}$, respectively. Otherwise, the element is 0. CNNs are adopted to explore the spatial correlation of the inputs and reduce the input dimension. The outputs of LSTM and CNN layers are concatenated together with the rest of the state and connected to fully connected layers. The output is the Q value or the WI value of the input state.

Algorithm 12 WI-based content delivery scheme

1: Initialize the weight vectors of neural networks θ_W and θ_Q for all user HMDs.
2: Initialize the exploration probability $\epsilon = 1$.
3: **while** time slot $n < n_{max}$ **do**
    % Schedule the content delivery requests.
4:   **for** user HMD $u = 1 : U$ **do**
5:      Request new video chunks $(i, s)_{u,n}$ through prediction, and report the WI $\lambda^*(Y_n^u; \theta_W)$.
6:   **end for**
7:   With probability ϵ, edge server schedules a request randomly.
8:   Otherwise, edge server schedules the request with the highest WI.
9:   The scheduled user HMD u receives the requested video chunks and updates the corresponding reward $R(Y_n^u, 1)$.
10:   Update slot $n = n + D(n, u)$.
    % Train the neural networks.
11:   **for** user HMD $u = 1 : U$ **do**
12:      Update the network for evaluating the Q value by minimizing loss function (5.28):

$$\theta_Q = \theta_Q - \varphi_Q \Delta L(\theta_Q). \qquad (5.31)$$

13:      Update the WI value by (5.29).
14:      Update the network for evaluating the WI by minimizing loss function (5.30):

$$\theta_W = \theta_W - \varphi_W \Delta L(\theta_W). \qquad (5.32)$$

15:   **end for**
16:   **If** $\epsilon > \epsilon_{min}, \epsilon = \beta_{attn}\epsilon.$
17: **end while**

To train the networks, as shown in Fig. 5.11, user HMDs obtain the instant reward according to (5.26). Parameters θ_Q in the Q value evaluation network are updated by minimizing the loss function in (5.28) according to the reward. Moreover, the Q value is further utilized to generate an approximated WI value by (5.29). The loss function (5.30) is to minimize the gap between the WI value approximated by the Q value and that approximated by the WI evaluation network. Parameters θ_W in the WI value evaluation network are updated accordingly. The accuracy of WI value approximation can be improved by learning the network dynamics and training the neural networks consecutively.

The proposed content delivery scheduling scheme allows user HMDs to train and infer their WI values in a distributed manner. The edge server makes scheduling decisions that adapt to network dynamics only by the collected WI values. The information on local devices, such as viewpoint trajectory and channel conditions, is not required by the edge server. By the proposed method, user HMDs infer the WI values in each time slot to measure the value of scheduling their requests. The edge server collects the WI values from the user HMDs and satisfies the request with the highest WI without requiring more information from the user HMDs, such as previous viewpoint trajectories. The information overhead can be significantly reduced by the proposed scheme. For example, let a value in the state be a float number (4 bytes). Only updating the previous viewpoint trajectory information, as a part of information in a state, requires a user HMD sending 400 bytes (2 float numbers for a viewpoint location) in each time slot if the viewpoint prediction window W is 50. Therefore, if the length of a time slot is 32 ms, at least 735 KB of data will be uploaded by a user HMD for playing a one-minute video. In contrast, in our proposed scheme, we evaluate a value for a user HMD state, i.e., WI. In each time slot, a user HMD uploads a float number (4 bytes) rather than the full state to the edge server for scheduling, which can significantly reduce communication overhead. Moreover, since the edge server only compares the WI values uploaded by user HMDs, the WI values can be further processed to reduce communication overhead, e.g., through quantization.

Furthermore, besides the decentralized structure for decision making, the proposed learning-assisted WI-based scheme can also schedule content delivery requests flexibly. First, the proposed scheme does not require a fixed number of VR users. VR users can enter or leave the system at any time slot, and the neural networks do not need to retrain their parameters due to environmental dynamics. Second, the proposed deep reinforcement learning provides a reference for scheduling rather than scheduling results. Thus, as an extension, it is possible to provide a priority-aware content distribution scheme by increasing the WI values for the user HMDs with high priority.

5.5.6   Numerical Results

In this subsection, we present the simulation results of the proposed AI-assisted mobile VR video delivery scheme.

Content Distribution Setting There are 12 users within the coverage of the edge server and watching the VR videos. The probabilities of any user watching video 1 to 3 are 0.3, 0.3, and 0.4, respectively. VSs in the same videos have the same VS popularity. Once a video is selected by a user, the user plays the whole video without interruption. The viewpoint movement profile of a user is selected randomly from the viewpoint movement profile among the 48 users in the data set. The transmission rates $R_{E,H}$ and $R_{E,L}$ are 250 Mb/s and 100Mb/s, respectively. The transition probabilities p_H and p_L are 0.6 and 0.3, respectively. The VR videos play at the rate of 30 frames per second. Thus, the length of a time slot is 32 ms, i.e., $1/g$, and a time segment, i.e., the time length of a VS, contains 125 time slots. The prediction window length W is 50 time slots. In each time slot, user HMDs only request the content corresponding to the current VS and the next VS. The video chunks of the VSs after the next VS will not be requested due to a likely high viewpoint prediction error.

Neural Network Setting Following the neural network structure shown in Figs. 5.10 and 5.11, there are three neural networks used for making content distribution decisions. The first is the neural network for viewpoint prediction. We apply one LSTM layer with 50 neurons to predict the viewpoint location in

Table 5.2 Network structure
for evaluating WI value

Layers	Number of neurons	Activation function
CONV1	$(3\times3\times 10)$	elu
POOL1	(2×2)	elu
CONV2	$(3\times3\times10)$	elu
LSTM	50	tanh
Fully connected 1	125	elu
Fully connected 2	64	elu
Fully connected 3	64	elu
Fully connected 4	24	elu
Output	1	Linear

Table 5.3 Network structure
for evaluating Q value

Layers	Number of neurons	Activation function
CONV1	$(3\times3\times 10)$	elu
POOL1	(2×2)	elu
CONV2	$(3\times3\times10)$	elu
LSTM	50	tanh
Fully connected 1	125	elu
Fully connected 2	64	elu
Output	1	Linear

the subsequent time slots. The prediction accuracy is around 94%. The other two neural networks approximate the Q value and WI value in Algorithm 12. Note that, besides state Y_n^u, we also utilize the statistical viewpoint distribution $p_{i,s}^P$ as a part of the input to improve the performance. The parameters of the neural networks for evaluating the Q value and the WI value are presented in Tables 5.2 and 5.3, respectively. The learning rates of the networks for evaluating the Q value and the WI value are 5e−4 and 1e−4, respectively. The initial exploration probability is 1, while the probability decreases with rate $\beta_{attn} = 0.9995$ after each iteration. The minimum exploration probability is 0.008.

Performance of the Content Distribution Scheme The performance of the proposed WI-based scheme is compared with three benchmark schemes: *urgent-request-first*, *round-robin*, and *random*. The urgent-request-first scheme always schedules requests with the minimum Δ_u. The round-robin scheme schedules requests in a sequential and rotational manner. The random scheme schedules requests randomly using a uniform distribution. We generate two different video selection and viewpoint movement profiles: one for training the neural networks, and the other for testing the learning-based scheme performance. For each profile, we generate requests for 2000 s. During the test, our scheme explores the environment with a minimum exploration rate and trains the neural networks. With the viewpoint movement prediction by AI-based method, the performance of the average hit probability in the four schemes is shown in Fig. 5.12, in which the

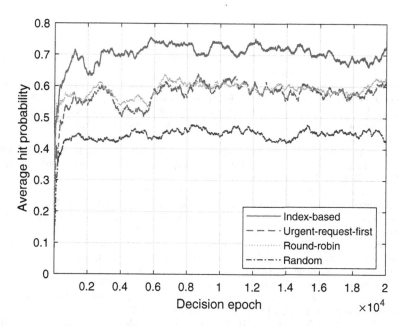

Fig. 5.12 Average hit probability with AI-based prediction and dynamic channel

test profile is applied. The results in Fig. 5.12 are the moving average of the hit probability in the past 1,000 decision epochs (around 8000 time slots) for each decision epoch. As shown in the figure, the proposed WI-based scheme can improve the hit probability by around 15% compared to urgent-request-first and round-robin schemes. The improvement is because, compared to the benchmark schemes, our proposed WI-based scheme evaluates the value for scheduling while considering the long-term rewards according to the states of user HMDs.

5.6 Summary

In this chapter, we have investigated content caching and distribution for mobile VR video streaming. Specifically, we have proposed a content placement scheme to cache popular and high-quality video chunks. We have also developed a novel learning-based content distribution scheme to schedule the video delivery for VR users. Taking advantage of the characteristics of VR videos, our proposed schemes can provide a scalable caching policy for a large amount of content with diversified content size and delivery delay requirements. The content distribution scheme proactively schedules video chunks in a low-overhead and flexible manner. The simulation results demonstrate that the proposed solution outperforms benchmark content caching and distribution policies.

References

1. Saad, W., Bennis, M., Chen, M.: A vision of 6G wireless systems: applications, trends, technologies, and open research problems. IEEE Netw. **34**(3), 134–142 (2020)
2. Maniotis, P., Bourtsoulatze, E., Thomos, N.: Tile-based joint caching and delivery of 360° videos in heterogeneous networks. IEEE Trans. Multimedia **22**(9), 2382–2395 (2020)
3. Boos, K., Chu, D., Cuervo, E.: FlashBack: immersive virtual reality on mobile devices via rendering memoization. In: Proc. 14th Annu. Int. Conf. Mobile Syst., Appl., and Services, pp. 291–304 (2016)
4. Du, J., Yu, F.R., Lu, G., Wang, J., Jiang, J., Chu, X.: MEC-assisted immersive VR video streaming over terahertz wireless networks: a deep reinforcement learning approach. IEEE Internet Things J. **7**(10), 9517–9529 (2020)
5. Dai, J., Zhang, Z., Mao, S., Liu, D.: A view synthesis-based 360° VR caching system over MEC-enabled C-RAN. IEEE Tran. Circ. Syst. Video Technol. **30**(10), 3843–3855 (2020)
6. Shi, S., Gupta, V., Hwang, M., Jana, R.: Mobile VR on edge cloud: a latency-driven design. In: Proc. 10th ACM Multimedia Syst. Conf., pp. 222–231 (2019)
7. Hooft, J.V., Vega, M.T., Petrangeli, S., Wauters, T., Turck, F.D.: Tile-based adaptive streaming for virtual reality video. ACM Trans. Multimedia Comput. Commun. Appl. **15**(4) (2019)
8. Sun, L., Mao, Y., Zong, T., Liu, Y., Wang, Y.: Flocking-based live streaming of 360-degree video. In: Proc. 11th ACM Multimedia Syst. Conf., pp. 26–37 (2020)
9. Zink, M., Sitaraman, R., Nahrstedt, K.: Scalable 360° video stream delivery: challenges, solutions, and opportunities. Proc. IEEE **107**(4), 639–650 (2019)
10. Sun, Y., Chen, Z., Tao, M., Liu, H.: Communications, caching, and computing for mobile virtual reality: Modeling and tradeoff. IEEE Trans. Commun. **67**(11), 7573–7586 (2019)
11. Wu, C., Tan, Z., Wang, Z., Yang, S.: A dataset for exploring user behaviors in VR spherical video streaming. In: Proc. 8th ACM Multimedia Syst. Conf., pp. 193–198 (2017)
12. Chakareski, J.: Viewport-adaptive scalable multi-user virtual reality mobile-edge streaming. IEEE Trans. Image Process. **29**, 6330–6342 (2020)
13. Perfecto, C., Elbamby, M.S., Ser, J.D., Bennis, M.: Taming the latency in multi-user VR 360°: A QoE-aware deep learning-aided multicast framework. IEEE Trans. Commun. **68**(4), 2491–2508 (2020)
14. Xiao, M., Zhou, C., Liu, Y., Chen, S.: OpTile: Toward optimal tiling in 360-degree video streaming. In: Proc. 25th ACM Int. Conf. Multimedia, pp. 708–716 (2017)
15. Li, C., Zhang, W., Liu, Y., Wang, Y.: Very long term field of view prediction for 360-degree video streaming. In: 2019 IEEE Conf. Multimedia Inf. Process. Retrieval (MIPR), pp. 297–302 (2019)
16. Xu, Y., Dong, Y., Wu, J., Sun, Z., Shi, Z., Yu, J., Gao, S.: Gaze prediction in dynamic 360° immersive videos. In: 2018 IEEE/CVF Conf. Comput. Vision Pattern Recognition, pp. 5333–5342 (2018)
17. Nasrabadi, A., Samiei, A., Prakash, R.: Viewport prediction for 360° videos: A clustering approach. In: Proc. 30th ACM Workshop Netw. Operating Syst. Support for Digital Audio Video, pp. 34–39 (2020)
18. Liu, Y., Liu, J., Argyriou, A., Ci, S.: MEC-assisted panoramic VR video streaming over millimeter wave mobile networks. IEEE Trans. Multimedia **21**(5), 1302–1316 (2019)
19. Mangiante, S., Klas, G., Navon, A., GuanHua, Z., Ran, J., Silva, M.D.: VR is on the edge: How to deliver 360° videos in mobile networks. In: Proc. Workshop Virtual Reality and Augmented Reality Netw., pp. 30–35 (2017)
20. Fehn, C.: Depth-image-based rendering (DIBR), compression, and transmission for a new approach on 3D-TV. In: Stereoscopic Displays and Virtual Reality Systems XI, vol. 5291, pp. 93–104 (2004)
21. Zhou, C., Li, Z., Liu, Y.: A measurement study of Oculus 360 degree video streaming. In: Proc. 8th ACM Multimedia Syst. Conf., pp. 27–37 (2017)

22. Petrangeli, S., Swaminathan, V., Hosseini, M., Turck, F.D.: An HTTP/2-based adaptive streaming framework for 360° virtual reality videos. In: Proc. 25th ACM Int. Conf. Multimedia, pp. 306–314 (2017)
23. Khuller, S., Moss, A., Naor, J.: The budgeted maximum coverage problem. Inf. Process. Lett. 70(1), 39 – 45 (1999)
24. Shanmugam, K., Golrezaei, N., Dimakis, A.G., Molisch, A.F., Caire, G.: FemtoCaching: wireless content delivery through distributed caching helpers. IEEE Trans. Inf. Theory 59(12), 8402–8413 (2013)
25. França, G., Robinson, D.P., Vidal, R.: A nonsmooth dynamical systems perspective on accelerated extensions of ADMM (2018). Preprint. arXiv:1808.04048
26. Whittle, P.: Restless bandits: activity allocation in a changing world. J. Appl. Probab. 25, 287–298 (1988)
27. Avrachenkov, K., Borkar, V.S.: Whittle index based Q-learning for restless bandits with average reward (2020). Preprint. arXiv:2004.14427

Chapter 6
Conclusions

In this chapter, we summarize the research in the preceding chapters of this book and the lessons learned. Then, we briefly discuss some future directions for connectivity and edge computing in IoT.

6.1 Summary of the Research

IoT has the potential to revolutionize human society, change the daily lives of people for the better, and create a smart and connected world in the future. While IoT use cases, such as smart homes, have been commercialized and gained popularity, we are still at an early stage of IoT research, development, and deployment. Significant efforts are necessary to propel the IoT towards the vision of connecting billions of IoT devices with on-demand data collection and analysis. Compared to the Internet, which revolutionized the world in the twentieth century, the IoT is much broader and supports many drastically different use cases. As a result, there is no "one-size-fits-all" solution, since IoT solutions must be customized for specific use cases. The technical content of this book has demonstrated the necessity and approaches of customizing connectivity and edge computing solutions in various representative IoT use cases.

For the smart factory use case in Chap. 2, simultaneously supporting a massive number of devices and guaranteeing very low communication delay is the main challenge, which is increased when communication overhead, device diversity, reliability, and complexity are considered. As a result, no existing MAC design can achieve all desired performance targets, and the proposed MAC protocol is designed to bridge the gap. Our customized design adopts the approach of centralized scheduling and distributed access, with a new slot structure, novel synchronization schemes, and a DNN assisted protocol parameter configuration, to maximize channel utilization, reduce delay, and guarantee low communication overhead. The resulting design can simultaneously satisfy the density requirement of mMTC and the delay requirement of URLLC for high-priority devices. A lesson from this chapter is the importance and difficulty of balancing different performance targets, such as device density, collision probability, delay, fairness, energy efficiency, and overhead. This also confirms the necessity of customized connectivity solutions, since the performance targets can be very different for different IoT applications such that making proper trade-offs is essential.

For the rural IoT use case considered in Chap. 3, a UAV acts as both a mobile AP and an MEC server. Contrary to the IIoT network scenario in Chap. 2, devices to be connected are dispersed over a large area relative to the coverage of the AP. Therefore, the main challenge is covering and serving more devices given a limited battery of the UAV, and energy efficiency becomes the main metric. Our approach to improving energy efficiency is to jointly optimize the UAV trajectory, which affects the connectivity between the UAV and devices, and the computing task allocation, which affects the edge computing services provided to connected devices. A lesson from this chapter is how the service provisioning cost, in terms of the UAV energy consumption, can have a major impact on the connectivity and computing in rural IoT applications.

For the IoV use case in Chap. 4, the main challenge is to guarantee computing service continuity in presence of vehicle mobility. Fortunately, different from the rural IoT scenario in Chap. 3, infrastructure including APs and MEC servers are usually available for vehicles in an urban or highway scenario. Therefore, we coordinate multiple neighboring MEC servers to guarantee service continuity and reduce service delay. The resulting problem involves server selection, computing load allocation, and result delivery, which can be too complex to solve using conventional optimization methods. We develop a deep reinforcement learning based approach to solve this problem, which demonstrates satisfactory adaptivity to vehicle mobility. A lesson from this chapter is the importance of judiciously designed coordination across multiple network segments, as well as the corresponding communication and computing resource scheduling for enabling such coordination, while handling IoT applications with rapidly changing network environments.

For the mobile VR video streaming use case in Chap. 5, balancing between the quality and quantity of cached video content is the key to improving the quality of experience for VR users. Different from the scenario in Chap. 3, in which the AP/server is mobile, and that in Chap. 4, in which the IoT devices are mobile, mobility is not a major concern. However, the viewpoints of VR users can be time-variant, which yields complexity due to the dynamics in VR content demands. Making the right trade-offs between caching and computing and between caching high-quality content and caching more content affects the users' quality of experience and is crucial to the application. A lesson from this chapter is the coupling relation between communication, caching, and computing in such IoT applications and how well-designed solutions should achieve the right trade-off based on the service demand, resource availability, and network dynamics.

The research works in the above four chapters serve a collective purpose: enabling on-demand data collection and/or analysis for IoT applications to improve various aspects of human society. The improvement made possible from our designs may manifest through higher productivity in factories, better connected rural areas, enhanced safety or convenience for vehicles, or a more comfortable or enjoyable entertainment experience. Of course, there are many more applications to consider, and many potential research directions to explore. We briefly discuss some of the directions next.

6.2 Discussion of Future Directions

There are many potential future directions for IoT research. Here, we elaborate on three promising directions, as follows.

The first is developing highly configurable connectivity solutions. The number of connected devices, the data traffic volume, and the connectivity requirements of IoT networks may change over time. For example, the number of connections in a vehicular network can vary significantly from peak to off-peak hours, and the type of connected devices in an IIoT network can be different in daytime and nighttime. Similarly, the number of connected devices on a farm may increase over time with the advancement of smart agriculture. To cope with the above changes, highly configurable connectivity solutions, such as configurable protocols, are desirable. Configurations may apply to the network organization, protocol adaption, access priority, communication pattern, and so on, while the configurations may be determined by AI. The objective of designing highly configurable connectivity solutions is to allow flexible on-demand trade-offs among connection density, spectrum efficiency, energy efficiency, delay, reliability, etc., since such trade-offs can be crucial to cost-effective operations of resource-limited IoT networks.

The second is creating intelligent and modular edge computing paradigms. Different IoT networks can have very different computing demands or computing capabilities, yet the basic components of edge computing are similar. The components may include computing task division, data uploading, computing task migration, computing task execution, collaborative computing, and result delivery. Accordingly, various modules can be created to manage the components, and a paradigm can be built from the modules to facilitate edge computing. Each module can integrate a set of candidate AI methods for achieving certain performance targets, and the paradigm can select and configure the modules based on the characteristics of the IoT application and the specific network setup. The objective of creating intelligent and modular edge computing paradigms is to improve the scalability of edge computing and provide a universal reference design of edge computing for various IoT use cases.

The third is designing data-centric IoT frameworks. Since data collection or exchange and data analysis are the core of IoT, the importance of data is evident. In addition to data required by IoT use cases, e.g., sensor readings for factory automation, data describing IoT devices and IoT networks, including data traffic volume, device mobility, resource utilization, service demands, and network performance, are also essential. Exploiting such data with proper AI methods can infuse intelligence into advanced network management for IoT use cases. Therefore, a data-centric IoT framework should consist of mechanisms to collect, store, update, organize, share, and process the data that describe IoT devices and networks, which should enhance the capability of networks to support corresponding IoT use cases. The mechanisms should also account for available network resources and application requirements. The objective of designing data-centric IoT frameworks is to achieve automated and data-driven IoT network management.

Index

© The Author(s), under exclusive license to Springer Nature Switzerland AG 2021
J. Gao et al., *Connectivity and Edge Computing in IoT: Customized Designs and
AI-based Solutions*, Wireless Networks,
https://doi.org/10.1007/978-3-030-88743-8

Printed in the United States
by Baker & Taylor Publisher Services